Space Power Interests

Space Power Interests

EDITED BY

Peter Hayes

Routledge
Taylor & Francis Group

LONDON AND NEW YORK

First published 1996 by Westview Press

Published 2019 by Routledge
52 Vanderbilt Avenue, New York, NY 10017
2 Park Square, Milton Park, Abingdon, Oxon OX14 4RN

Routledge is an imprint of the Taylor & Francis Group, an informa business

A CIP catalog record for this book is available from the Library of Congress.

ISBN 13: 978-0-367-28850-1 (hbk)
ISBN 13: 978-0-367-30396-9 (pbk)

Contents

Introduction

Peter Hayes

In this book, we ask the simple question: can the world avoid the proliferation of long range missiles in forthcoming decades? In part, this question arises because many medium and small states are industrializing rapidly, implying the equally rapid diffusion of basic technological capabilities required to exploit space--including booster rockets known as space launch vehicles (SLVs). In many respects, SLVs are identical to intercontinental ballistic missiles. Which states then might seek to acquire independent SLV capability, and of these states, which might have the requisite financial means and technological prowess to pursue this option?

Gaining an SLV capability, however, is only one route to achieving de facto intercontinental ballistic missile (ICBM) technological capability. Others--as shown in Table 1--are indigenous missile production, modification of purchased missiles, outright purchase of missiles, purchase of warheads avoiding need for a local development program, overlapping SLV and missile programs, and underlying each of these paths, the necessary financial wherewithal.

Five sets of actual or potential missile proliferating states may be identified by applying these missile proliferation technology paths to observed proliferation activity, as is shown in Table 2. (As is evident, there are no cases as of 1993 of states buying warheads of mass destruction; and, by definition, all these states found proliferation to be affordable. By implication, the same applies to states posited to proliferate long range missiles in 2000 and 2010.)

In Table 3, a scenario is shown in 2010 as to which pathways various states might have travelled to obtain long range missiles. In this view, the SLV route is likely to be a critical pathway for states pursuing a long range missile capability--especially in Asia with its many latecomers to industrialization--given the enhanced controls on the more direct routes to producing, modifying, or buying long range missiles, combined with the value of independent access to space for wealthy states. Yet of the seven prospective ICBM-capable states in this region, only two (China and Japan) are formally committed to the Missile Technology Control Regime (or MTCR), itself an informal arrangement.

1

Table 1 Routes to Long Range Ballistic Missile Capability

1. *Indigenous BM Production*
   Technological: guidance, control, propulsion, structures, materials
   Human resources: scientists and engineers
   Facilities: laboratories, ground and flight testing
2. *Modification of Purchased BMs*
   Technological: refit guidance and control systems; testing
3. *Purchase of BMs*
   Competent military
4. *Production or Purchase of Space Launch Vehicle*
   Technological: reentry vehicle, heat shields, tracking and control
5. *Warheads*
   Technological: biological, chemical, nuclear
6. *Overlapping, opaque SLV orbital launch and BM programs*
7. *Funding*
   Thresholds
   Military expenditure

Notes: BM = ballistic missile; SLV = space launch vehicle

Table 2   States Armed, Equipped, or Capable of Fielding Long Range (>1,000 km) Ballistic Missiles (1993, 2000,2010)

Route to BM Capability		Time	Armed, Equipped, Capable States
A.	Armed	1993	China, France, Russia, United Kingdom, United States
	Equipped	1993	Belarus, Khazakstan, Ukraine
B.	Indigenous	2000	Brazil, India, Germany, Italy, Israel, Japan, Sweden
	BM Prodiction	2010	Argentina, Israel, S. Korea, S. Africa, Taiwan
C.	Modification	2000	Iran, N. Korea, Pakistan
	of BM Imports	2010	Egypt, Libya, Syria
D.	Purchase	1993	Saudi Arabia
	of BM	2000	Argentina, Egypt, Iran, Iraq, Syria
E.	Hybrid of	1993	India, Israel, Japan
	BM routes	2000	Argentina, Brazil
	+ SLV Production	2010	Pakistan, S. Korea, S. Africa, Taiwan, Ukraine
	or Purchase		

Key: BM = ballistic missile; SLV = space launch vehicle Note: By 2000, North and South Korea may be reunified; by 2010, it is likely. Source: Proliferation Study Team, *The Emerging Ballistic Missile Threat to the United States*, report to US Department of Defense, February 1993.

Unfortunately, some of these states are involved in regional conflicts and their elites have exhibited great power aspirations. Thus, there is a prima facie concern that the MTCR may be unable to constrain actual or latent ICBM proliferation.

Table 3 Ballistic Missile Candidate States, 2010

Long range (> 1000 km with 1 tonne payload); missile -armed, -equipped or -capable

States by Region		*Route*				
		A	B	C	D	E
South America						
	Argentina		*		*	*`
	Brazil		*			
Middle East						
	Egypt			*	*	
	Iran			*	*	
	Iraq				*	
	Israel		*			*
	Libya			*		
	Saudi Arabia				*	
	Syria			*	*	
South Asia						
	India		*			*
	Pakistan			*		*
East Asia						
	China	*				
	Japan		*			*
	North Korea			*		
	South Korea		*			*
	Taiwan		*			*
Africa						
	South Africa		*			*
Europe						
	France	*				
	Germany		*			
	Italy		*			
	Sweden		*			
	United Kingdom	*				
Former Soviet Union						
	Russia	*				
	Belarus	*				
	Khazakstan	*				
	Ukraine	*				
North America						
	United States	*				

Key--Ballistic missile acquisition path: A: BM armed or equipped; B: Indigenous BM production; C: Modification of BM imports; D: Purchase of BM; E: Hybrid of BM routes and SLV production or purchase. Italicized states: 1993 MTCR participants or unilateral adherents. Source: Proliferation Study Team, *The Emerging Ballistic Missile Threat to the United States*, report to US Department of Defense, February 1993

Table 4 Financial Threshold for Nuclear-Armed, Long-Range, Missile-Capable States

Program

10-20 nuclear warheads per year over 5 years

Suite of short/medium range air and missile delivery systems, including IRBM/ICBM

Element of Cost	Estimated Annual Cost Billion US$	Estimated Five Year Cost: Billion US$
(a) Warhead RD@D: 10-20/year	0.2	1.0
(b) Acquisition of strike bombers and short range ballistic missiles	0.3	1.5
(c) Infrastructure, training, radar	0.2	1.0
(d) Indigenous BM > 1,000 km	0.4	1.0
Total	1.1	4.6
Threshold Program Expenditure	**1.0**	**5.0**

Sources: G. Rochlin, "The Development and Deployment of Nuclear Weapons Systems in a Proliferating World," in J. King *ed, International Political Effects of the Spread of Nuclear Weapons*, Central Intelligence Agency, April 1979, pp. 20-21; C. Wolf *et al, Long-Term Economic and Military Trends, 1950-2010*, Rand N-2757-USDP, April 1989; Stockholm International Peace Research Institute, *SIPRI Yearbook 1992*, Oxford University Press, New York, 1992, Table 7A2, pp. 259-263; S. Meyers, *The Dynamics of Nuclear Proliferation*, Chicago University Press, Chicago, 1984, pp. 38-40; Proliferation Study Team, *The Emerging Ballistic Missile Threat to the United States*, report to US Department of Defense, February 1993, pp. 7-8.

Potential proliferators face the high cost of a missile program, whichever route is taken. In Table 4, an annual financial threshold of about $1 billion investment is defined for a minimal, five year long warhead development and delivery program, of which about $0.4 billion is related to long range missile capability. Using this criteria, Table 5 identifies 24 states currently without long missiles which could "afford" such a program in the future (2010). If states must have both technological and financial capabilities to be long-range missile capable, as defined above, then one can narrow the field to the ten states shown in the left column of Table 6, mostly in the Middle East and Asia.

It is possible, therefore, to sketch the candidates for missile power status in 2010. In tier one are states that have already armed with long range missiles (United States, Russia, China, United Kingdom, France, Saudi Arabia, and India).

In tier two are fourteen states which have both the means and the capability to achieve this status by 2010, mostly in the Middle East and in Asia. In tier three are five states which by extraordinary effort could acquire (or, as with North Korea, already have acquired) long range missiles. However, they are likely to find it hard to translate future missile activity into strategic arsenals due to continuing resource constraints. In tier four are nine states which have the

Table 5 Candidate States for Long Range Ballistic Missile Status, 2010
Based on Threshold Expenditure

Criteria: annual $1 billion program expenditure cannot exceed 20 % of candidate state's annual military expenditure.
Minimum annual military expenditure to afford such a program is $5 billion.

A. Candidate states, 1991 annual military expenditure > $5 billion
Australia , *Canada, China, France* , *Germany,* India, Iran, Iraq, *Italy, Japan, Netherlands, Russia,* Saudi Arabia, South Korea, *Spain,* Taiwan, *United Kingdom, United States*

B. Additional Candidate states, 2010 annual military expenditure > $5 billion
(assumes 3.5% real annual growth in 1991 military expenditure)
Belgium, Egypt, *Greece, Israel, Norway,* Poland, South Africa, *Sweden, Switzerland*, Syria, Turkey

Bottom Line: 24 candidate states in 2010 over and above existing ICBM-armed states.
Note: Italicized states: 1993 MTCR-participants or unilateral adherents to the MCTR

requisite resources, but are likely to forego obtaining a missile capability whether via a missile or an SLV program. These states are all OECD states allied to the United States, or European states. Finally, there is a fifth tier of about 120 states which are projected to have neither the means nor the capability in 2010 to threaten long range missile proliferation. Within this group, states such as Indonesia and Vietnam may have grown sufficiently to be seriously exploring a missile option.

In this book, our concern is to explore the arms control and disarmament measures which apply primarily to the first three tiers on the assumption that tiers four and five are either already committed non-proliferators, or will simply lack the ability to enter the field. Thus, we focus on the proliferation propensity of ambivalent and hard core proliferators, and the interests of existing long range missile powers to modify or abandon their own capabilities to contain the proliferation of second and third tier states. In Table 7, I classify the latter, thirteen missile capable states into ambivalent versus hard core proliferators. A worst case profile of all states that could have long range missiles in 2010 is constituted by these thirteen states plus the seven states already so armed.

Undoubtedly, the proliferation threat exists. But obviously, this threat arises primarily from local and regional conflicts--albeit often linked to great power capabilities. Consequently, non proliferation measures must be tailored to specific regional problems and circumstances and will differ from state to state. Ambivalent proliferators--many of which are concerned about their reputation as "responsible" states--may respond to long range missile non proliferation regimes, provided that their fundamental security concerns are met and provided that adversarial great power or hard core proliferators do not threaten them directly. Ambivalent proliferators often rely on alliances to extend deterrence or defense against threats posed by hard core proliferators or great powers armed with missiles. For precisely this reason, hard core proliferators may reject

Table 6 States with Missile Capability and Financial Means to Proliferate, 1991-2010

	Technology *and* Finances	Technology Only	Finances Only
Europe			
	Germany		Greece
	Italy		Netherlands
	Sweden		Norway
			Poland
			Spain
			Switzerland
			Turkey
Middle East			
	Egypt	Libya	
	Iran		
	Iraq		
	Israel		
	Syria		
South Asia			
		Pakistan	
East Asia			
	Japan	North Korea	
	South Korea		
	Taiwan		
Africa			
	South Africa		
South America			
	Argentina		
	Brazil		
North America			
			Canada
Australasia			
			Australia

Note: States in the left column are projected to acquire both technological long range missile capability by 2010, and to meet the financial threshold criterion. States in the middle column are projected to acquire technological missile capability, but to not meet the financial threshold criterion. States in the right column are projected to meet the financial threshold criterion, but to not acquire technological missile capability. Also, existing long range missile armed states (United States, Russia, China, United Kingdom, France, India, Saudi Arabia) are not shown, and "temporary" missile-armed states (Belarus, Khazakstan, Ukraine) are assumed to disarm.

universal non proliferation regimes relating to long range missiles even if these offer substantial incentives for participation. Such delivery systems enable these states to threaten patron states of their local or regional adversaries. Thus, universal regimes for "Rockets for Peace" relate primarily to the interests of states committed to non proliferation (even if they are themselves already armed with long range missiles), and to states which are ambivalent about their non proliferation commitments. Per se, such regimes may hold little attraction for hard core proliferators.

Table 7 1993 Propensity to Proliferate versus 2101 LRBM-Capability

Propensity of State	Armed 1993	LRBM Capable 1993	LRBM Capable 2000	LRBM Capable 2010
Already Proliferated	United States Russia China France United Kingdom India Saudi Arabia			
Non Proliferators		Brazil	Argentina	Egypt South Africa South Korea Taiwan
Hard Core Proliferator		Iran	Israel Iraq Pakistan	Libya Syria North Korea

Note: By 2000, North and South Korea may be reunified; by 2010, it is likely that Korea will be unified.

Given this roughly mapped terrain, the authors of this book examine the notion that a "Rockets for Peace" regime is desirable and/or feasible. Given the dominance of their concerns in the international system, the first five chapters examine the interests of the existing space and missile powers in achieving non proliferation of long range missiles. In Chapter 1, Thomas Mahnken and Janne Nolan define the challenges posed by space launch and missile proliferation. In particular, they describe the motivations that may lead states to acquire ballistic missiles, assess the current missile and space programs of Asian nations, and analyze future trends in missile proliferation. They focus on the ability of developing states to convert space launch vehicles (SLVs) into long-range ballistic missiles.

They conclude that the relative weakness of the MTCR and the limited utility of traditional arms control measures for controlling the diffusion of missile technology imply that a "missile restraint regime" should be subsumed into ongoing efforts to end regional conflicts with an immediate focus on modest confidence-building. These confidence building measures could ease unwarranted suspicions about missile production efforts, limit their political and military consequences, and even reduce some of the motivations underlying missile proliferation.

In Chapter 2, John Pike and Eric Stambler examine the interests of the United States--the world's leading missile and space power--in promoting a new space access regime that would provide incentives to regime participants to not proliferate long range missiles. Pike and Stambler conclude that a necessary condition for establishing a successful long range missile non proliferation regime is that all states must share in the prestige of space if they are to be persuaded to forgo self-reliant capabilities. "This outcome," they argue, "can be accomplished by allowing access or association with the technology and activities of spacefaring nations." Once adopted, this new regime would eliminate the status of long range rockets as prestige weapons and replace it with cooperation in civil space exploration and development. In a prescient argument, they suggest that the latter may become a defining activity of the coming millennium -- of which the US-Russian space station may be the precursor.

In chapter 3, Maxim Tarasenko contends that Russia is and will remain an important space and missile power of central importance in any discussion of a long range missile non proliferation regime. Delving back into the missile and space programs of the former Soviet Union which still shape that of the new Russia, Tarasenko concludes that Russia would both win and lose from a regime that gives potential missile proliferators an incentive to find peaceful applications for space technologies in a new non-proliferation regime. He suggests that Russia might support strongly a practical version of this concept because it could enhance international security as well as offering Russia direct economic benefits via use of excessive capabilities, and, in the long run, by integrating the Russian rocket and space industry into the world system. Tarasenko argues that the most important outstanding issues are: (1) establishing the principle of access to space launch services; (2) defining the obligations of the parties and guarantees of compliance; and (3) negotiating a universal list of space activities that are allowed and prohibited for all countries.

In Chapter 4, Yanping Chen reviews China's space interests and missile tech-nology controls as embodied in the MTCR. Chen notes that space technology pro-ducts spun off from missiles constitute a large proportion of high technology products sold by China today. After describing China's SLVs and long range ballistic missiles, she analyzes Chinese motivations for developing rocket technology during the Cold War era as well as commercializing space products during the early eighties. She holds prestige to motivate primarily China's drive to acquire and sell these systems. She also concludes that China would likely accept international cooperation in various space programs in exchange for a reduction of the military use of space technology. Conversely, she believes that China is unlikely to commit itself to a more restrictive missile technology regime, especially one that might hurt its space industry.

In Chapter 5, Jurgen Scheffran shows that Europe's space program is undergoing a rapid and radical transformation in the aftermath of the Cold War and confronted with the realities of high cost and low returns. Scheffran observes that emerging missile threats from developing countries are not a high political priority in European politics, but this may change if European capitols were to come within flight distance of missiles from developing countries. The Europeans also recognize that the MTCR is discriminatory and prevents many cooperative space programs with developing countries. Therefore, he concludes, most European countries would probably favor demand-side oriented solutions, based on missile arms control and disarmament which would not block Europe's space efforts.   Verification that non proliferation commitments are being implemented would be critical if Europeans were to participate in such a scheme.

In the second section of the book, the authors examine the precedents and potential for a universal, long range missile non proliferation regime. In Chapter 6, Joan Johnson-Freese asks if increasing cooperative space activity might affect positively efforts to halt long range missile proliferation.   Relatedly, she considers whether some formal regime or even a World Space Agency for controlling the technology would be either desirable or feasible, based on past experience. She argues increased space cooperation does not increase the risk of missile technology proliferation, largely because rocket technology is not normally acquired cooperatively in any case.   Rather, developing rocket technology has been a highly nationalistic activity. For the same reason, there is not much foundation in common interest for institutionalized cooperation to emerge for missile non proliferation or indeed any other objectives related to space-based activities. Given strong perceived national interest, she concludes that a treaty or a formal organization reliant upon intrusive inspections to implement a cooperative space regime are unlikely to be well received by space-faring nations. In short, if a country seeks missile technology, then it will be able to obtain it whatever is done cooperatively in relation to space.

In Chapter 7, Molly Macauley examines the thesis that space-based economic or environmental benefits might be sufficiently valuable to motivate states to commit to a long range missile non proliferation regime. Macauley argues that linking environmental benefits to defense tradeoffs such as foregoing long range missiles is not feasible. In short, more direct approaches to non-proliferation--improved verification, challenge inspections, the sharing of intelligence--may be the more fruitful strategy.

In Chapter 8, Lucy Stojak reviews the experience with arms control and verification relating to long range missile non proliferation. Because ballistic missile and space-launch vehicle technologies are virtually identical, an international legal framework is needed to ensure that the exploration and use of outer space be carried out for the benefit and in the interests of all countries. Stojak asks how such a regime might be created for outer space which makes allowance for the dual-nature of space technologies while advancing the use of applications which promote stability? To this end, she reviews past proposals for on-site inspection, as well as institutional arrangements discussed in the United Nations and the Conference on Disarmament.

Stojak argues that the prospects for a multilateral agreement on dual-use technologies are poor. Consequently, she suggests that confidence building measures in outer space activities and the use of its technologies are more pragmatic and likely to be accepted. Relatedly, the MTCR suffers from a number of defects including the failure of major suppliers to join the regime, the growing sophistication of production capability in potential suppliers (which also have not joined the MTCR), the increased risk of proliferation stemming from the weakness of enforceable export controls in the states of the former Soviet Union, and, above all, the fundamental inability of any supply-side controls to halt proliferation. Thus, she concludes, the MTCR must evolve from an export control regime to a broader multilateral non-proliferation arrangement that develops and promotes international norms in the transfer and control of missile technology.

In Chapter 9, Lora Lumpe examines the argument that a flight test ban could be an effective tool to curb long range ballistic missile proliferation. She notes that this idea has been considered during decades of superpower arms control but never implemented. In the current context, she asks whether a flight test ban would limit effectively ballistic missile development and if it could be verified given the issue of SLV flights. She concludes that a global and total flight test ban would freeze existing ballistic missile developments, and gradually erode existing stockpiles. Although special verification measures would be needed, a flight test ban treaty would not be very difficult to verify via satellite and aerial reconnaissance. Politically, a flight test ban could be a useful way to demonstrate the commitment of existing nuclear and long range missile powers to nuclear arms reduction.

In Chapter 10, Timothy McCarthy draws out some lessons for ground-based verification of a test ban treaty from the UN Special Commission on Iraq (UNCSCOM) experience in dealing with Iraq's weapons of mass destruction, including its missiles. Via his organizational analysis of UNSCOM's Ongoing Monitoring and Verification (OMV) of Iraq (which uses intrusive on-site inspection), McCarthy shows how it has established new techniques for controlling weapons proliferation. He concludes, however, that UNSCOM is unique and unlikely to be replicated in other hard core or ambivalent proliferating states. Rather, McCarthy suggests that UNCSCOM provides four crucial lessons, namely, that: (1) a small, efficient and non-bureaucratic monitoring and verification organization may be the most effective; (2) it requires UN Security Council support to be effective; (3) a UN intelligence agency is feasible; and (4) an effective agency faced with an uncooperative adversary can do the job, even in a politicized context like the United Nations.

In Chapter 11, Peter Zimmerman addresses the technical challenges posed by the verification of ballistic missile activities. He argues that new proliferating states may be satisfied with low-end but credible weapons of mass destruction, at least at the outset. Thus, supply-side restrictions on "bronze medal" technologies discriminate between nations, are difficult to impose, and are likely to fail in the long run. Zimmerman shows that if a treaty could be achieved to control long range (greater than 300 km) missiles, then there are no insurmountable technical barriers to distinguishing between peaceful space launches and activities versus military-related testing and deployment of rockets. Such surveillance can be conducted on a cooperative and non-intrusive basis and could support a universal or regional regime controlling missile tests and deployments.

In short, the authors show that universal and regional approaches are feasible and badly needed to strengthen the existing ballistic missile non proliferation regime. Faced with hard core missile proliferators, special steps involving the UN Security Council may be required to impose controls rather than to induce compliance with missile non proliferation regimes. But some ambivalent missile proliferators may find regional approaches such as missile test ban treaties tailored to their specific insecurities to be more attractive than gaining independent missile forces.

Undoubtedly, the link remains between the demonstration effect and military utility of the standing missile forces of existing missile powers and the motivation of ambivalent and hard core proliferating states. A precondition of long run successful missile non proliferation, therefore, seems to be that the United States and Russia continue to reduce their long range missile forces on the one hand; and that other states armed with long range missiles accept controls on and start to reduce their missile forces on the other. Perhaps the missile test ban treaty is the missing link between these two categories of missile proliferation. What remains unclear is to what extent tangible or symbolically significant space-based services or access to space might induce states to participate in a "rockets for peace" regime.

The editors would like to thank all the authors for responding to a gruelling review process. We also thank the staff of Monterey Institute for International Studies for their assistance in hosting a workshop based on earlier versions of these papers. Finally, we thank the Alton Jones, John Merck, Ploughshares, and Winston foundations for their financial support for the research included in this volume. Everyone associated with this effort would like to thank Ms. Paula Fomby, formerly of the Nautilus Institute, for her unfailing attention to detail and effort in bringing the project to timely completion.

1

Challenges Posed by Space-Launch and Missile Proliferation

Thomas G. Mahnken
Janne E. Nolan

In this chapter we describe the motivations that may lead states to acquire ballistic missiles, assess the current missile and space programs of Asian nations, and analyze future trends in missile proliferation.[1] We focus on the ability of developing states to convert space launch vehicles (SLVs) into long-range ballistic missiles. We conclude by examining whether the present ballistic missile nonproliferation regime can contain adequately the spread of missile technologies without imposing excessive burdens on desirable space investment.

Motivations

Developing states seek ballistic missile and SLV capabilities for a host of reasons; in most cases it is difficult to isolate a single overriding motivation. It is difficult to formulate policies to address the demand for such systems without a better understanding of the motivations that lead states to acquire ballistic missiles and the interplay among them. Single-factor explanations can be especially deceptive. An economic analysis may indicate that third world SLV programs are unlikely to be profitable, for example. However, such single-factor analysis does not capture other, equally compelling motivations, including the desire of developing states to pursue such programs for reasons of political prestige and self-sufficiency, even if they prove uneconomical.[2]

Let us first address political motivations for developing ballistic missiles and SLVs. Ballistic missiles (and space systems in general) are symbols of national power and prestige in many developing states. Such programs often inspire patriotic and nationalistic sentiment, demonstrating both technical capacity and industrial sovereignty. In this respect, missile technology today holds the same kind of symbolic importance as the dreadnought at the beginning of this century: a physical demonstration of a nation's entry into a position of global importance. As Indian strategic analyst K. Subrahmanyam wrote after the first test of the Indian Agni intermediate-range ballistic missile (IRBM) in 1989, "[Its] role as a weapon is the least of its roles. It is a confidence builder and a symbol of India's assertion of self-reliance not merely in defense but in the broader international political arena as well."[3]

Some states also build an aerospace production base to generate revenue through arms exports and to expand their scientific and technical infrastructure. Brazil and North Korea stand out as states which were motivated to produce missiles largely for profit, although Brazil had to confront the quixotic nature of the undertaking and restrict its missile programs in the early 1990s. The development of an indigenous manufacturing base for ballistic missiles may prove to be an attractive alternative to the production of more complex weapons systems, such as aircraft. In the future, the sale of turnkey missile factories may provide less well-developed states with capabilities they would be unable to achieve on their own. Investment in a missile production program may also serve as a first step toward the development of a more sophisticated aerospace industry. China, for example, used technology from its ballistic missile program to nurture its drive for space exploration.

Conversely, by developing an indigenous space launch capability, states may hope to reap the benefits of space technology while also retaining an option to deploy ballistic missiles. The transfer of space technology may also forge new transnational relationships. For example, Brazil and China have forged an agreement to deploy a constellation of earth-imaging satellites and market ground stations and space data.[4]

But missiles are more than mere symbols of national power. They are also military instruments of potentially strategic significance. On a tactical level, ballistic missiles allow states to strike distant targets quickly and with little or no warning, and do not yet confront effective defenses. They are also immune to pilot error and require limited logistical and manpower support. Partly for these operational reasons, ballistic missiles are the weapons of choice of many states in the developing world.

Ballistic missiles also can be attractive complements to or substitutes for air forces for striking enemy territory. Recent analysis has attempted to examine the relative effectiveness of ballistic missiles and strike aircraft for delivering a payload, concluding that aircraft are far more capable.[5] However, such an approach is simplistic and may even be irrelevant, in part because most developing states do not face such a tradeoff. Air forces in the developing world are largely tactical, dedicated to providing air defense of the homeland and close air support to forces in the field rather than conducting strategic bombing missions. Such analysis also ignores the potential synergy between aircraft and ballistic missiles, which can serve as an effective force multiplier. For example, ballistic missiles could be used to strike air bases to clear corridors for TMD attacking planes, while aircraft could be used to deliver anti-radiation missiles to suppress theater missile defense systems.

The use of ballistic missiles has become a hallmark of conflict in the Third World. The systems employed to date, such as the modified Scud missiles launched by Iraq against Israel and Saudi Arabia during the Gulf War, are still highly inaccurate and carry relatively small high-explosive warheads. But they had a significant impact on civilian morale (during the "War of the Cities" in the Iran-Iraq War) and on military operations (during the Gulf War). Moreover, a new generation of missiles with greater range, higher accuracy, and with the capacity to carry more destructive payloads is appearing on the world market. These systems will pose a much greater potential threat to regional stability than did Saddam Hussein's missiles.

At the strategic level, ballistic missiles can serve as the means to exert coercive leverage against regional rivals or to deter US intervention in local conflicts. By possessing the ability to strike neighboring adversaries, states may hope to coerce the former into denying requests for access to the region by US forces. A couple of examples will serve to make the point. Had Iraq launched missiles against active airbases during the crucial buildup of coalition forces when fully-loaded transport aircraft were parked on the tarmac, then its missile strikes could have been much more destructive than those actually launched during the course of the war. As it was, the single, random Scud which hit a barracks at Dhahran killed 28 soldiers and wounded 100, which proved to be the highest level of casualties of any engagement in the war. On February 16, 1991, an Iraqi Scud struck the Saudi port of Jubayl barely 300 meters from a pier at which eight vessels were docked, among them two containing virtually all the provisions for the US Marine Corps air wings, several carrying ammunition, the USS Tarawa amphibious craft, and a Polish hospital ship. Fortunately the strike inflicted no damage, and even left untouched 5,000 tons of 155-mm artillery shells stacked on the pier that day.[6]

Future adversaries may hope to deter outside states from intervening in regional affairs. As Libyan leader Col. Khaddafi told one audience, referring to the 1986 air raid on Libya:

> Did not the Americans almost hit you...when you were asleep in your homes? If they knew that you have a deterrent force capable of hitting the United States, they would not be able to hit you. If we had possessed a deterrent--missiles that could reach New York--we would have hit it at the same moment. Consequently, we should build this force so that they and others will no longer think about an attack.[7]

Hyperbole aside, it appears that this kind of rationale is gaining ground elsewhere. Iranian analysts have expressed the need to deploy missile systems to protect the country against external--meaning Western--attack. Proponents have explicitly argued that Iran should "build up its own short, medium and long-range surface-to-surface as well as surface-to-air missiles" in order to "boost the defense capabilities of the country and minimize possible enemy air and missile strikes against economic centers as well as military forces."[8] A number of Indian defense analysts have made similar recommendations. According to Air Commodore Jasjit Singh, director of India's Institute for Defense Studies and Analyses, "Time has come for the Indian Air Force to think of integrating its deep strike aircraft with a possible ballistic missile force to constitute a strategic deterrent capability for national defense."[9]

Programs

In seeking a ballistic missile capability, developing states have pursued three alternative paths: indigenous development, modification of existing systems, and outright purchase. Although it is likely that indigenous development of ballistic missiles will be limited to states possessing a substantial science and technology infrastructure, the purchase and modification of systems provide viable options for a wide spectrum of states.

The time and resources required to develop ballistic missiles will be a function of the path chosen. Although the purchase of a handful of missiles and launchers may be relatively inexpensive relative to the acquisition of weapons such as strike aircraft, the development of the facilities needed to deploy and maintain an indigenous missile force would be a much more expensive and time-consuming undertaking. Industrialized states may be able to reduce the investment required by tapping domestic arms manufacturing capabilities and related aerospace, chemical and metallurgical industries to produce missiles.

To date, the vast majority of ballistic missiles in the developing world have been purchased from a few suppliers, including the former Soviet Union, China, and North Korea. Despite technology-transfer controls, industrial states are likely to continue to be a major source of ballistic missile diffusion. Although the existence of a 22-nation cartel limiting transfers of missile technology--the Missile Technology Control Regime (MTCR)--has reduced the ranks of states willing to export ballistic missiles, a number of countries may continue to sell missiles and associated technology both to increase their influence in regional affairs and to generate hard-currency earnings. Most missiles exported to date have been liquid-fueled, relatively short-range, inaccurate, and equipped with small high-explosive warheads. In the future, however, producers are likely to diversify the types of systems offered for sale, including missiles with longer range, greater accuracy, and more lethal warheads. The Chinese M-9 appears to be one example of an attempt to meet the demands of foreign customers for a missile superior to the Scud B.[10] North Korea has reportedly offered the 1,000-km No Dong I for sale, and may be developing an even longer-range missile for export.[11] The acquisition of production facilities appears to be an increasingly attractive option for developing states who hope to minimize their dependence on outside suppliers. Egypt, Syria and Iran for example either have received or are in the process of acquiring turnkey factories to produce ballistic missiles.[12]

At the other end of the technological spectrum, a number of states, including Brazil, China, India and Israel, have pursued indigenous efforts to develop ballistic missiles for decades, several involving the development of missiles with a range of 1,000 km or more. These have been developed both for security purposes and for export. Even for states with this level of development, the creation of an indigenous missile infrastructure can pose a daunting challenge. It requires a cadre of scientists and engineers capable of designing, testing and manufacturing missile systems, as well as a technical infrastructure capable of supporting such a program. As Brazil's recent decision to forgo its missile programs demonstrates, indigenous development will likely remain an attractive option for only more technologically advanced states.

Given these difficulties, a number of states have chosen a middle path, modifying ballistic missiles purchased from suppliers to increase their performance. This is the option Iraq chose when it extended the range of Soviet-supplied Scud Bs to produce the Al-Hussein and Al-Abbas missiles. North Korea, similarly, has produced an extended-range (500-km) Scud, dubbed the Scud C, which is believed to have been sold to Iran and Syria.[13] Modification can be a first step towards achieving a more capable missile production sector. China, India, and South Korea modified surface-to-air missiles to produce ballistic missiles.[14] Pakistan, for its part, converted sounding rockets to produce the Hatf surface-to-surface missiles,[15] while a number of other states have converted space launch vehicles for use as ballistic missiles.

As missile proliferation becomes an increasingly important security challenge, one key area to examine is the potential for converting civilian space launch vehicles for use as surface-to-surface missiles.[16] The link between space launch vehicles and surface-to-surface missiles is a strong one: the United States, Soviet Union, China and Israel have all used ballistic missiles as space launchers. India's Agni IRBM utilizes a first stage derived from its SLV-3 space launch vehicle, while Brazil converted its Sonda series of sounding rockets into artillery rockets.[17] The MTCR includes sounding rockets and space launch vehicles and their components as items to be restricted by the regime. In addition, the United States has barred exports to a number of third world SLV programs, including those of India, Brazil, and South Africa, on the grounds that such technology could aid in the development of surface-to-surface missiles. It is too early, however, to evaluate the long-term effectiveness of such sanctions.

Developing states seek an indigenous space launch capability for a variety of motives, ranging from the desire for technological prestige and independence to the prospect of export earnings through the sale of launch services. The development by third world states of SLVs has often been an integral part of programs to exploit space for a range of civilian development activities, including communications, environmental monitoring, and earth observation. The development of SLVs by third world states also provides a latent capability to develop long-range ballistic missiles. Ballistic missile development programs launched under the auspices of a space program may find it easier to attract greater foreign technical assistance than would an overtly military project. France, for example, assisted space programs in Brazil and India, while Russia sold cryogenic rocket engine technology to India, ostensibly for civilian purposes.[18] The mere existence of a space launch vehicle program does not in itself indicate the intent to produce an intercontinental ballistic missile (ICBM), of course, but the existence of an operational SLV capability does allow a state to develop a long-range ballistic-missile capability relatively quickly, should it so desire.

For example, China's space program was adapted from Soviet technology by Dr. Chien Hsu-shen, a research engineer at the California Institute of Technology who was expelled from the United States in 1955 during the U.S. anti-Communist campaigns.[19] The Chinese currently offer commercial launch services with their Long March family of launchers, a number of which are derived from China's intermediate-range and intercontinental ballistic missiles. According to press reports, China has transferred or plans to transfer missile technology to Iran, Pakistan, Syria and Saudi Arabia. In addition, China has signed space technology cooperation agreements with Brazil, Israel, and South Korea. Although China agreed in 1991 to abide by the MTCR, it is not clear that it will enforce export restrictions given competing economic and politico-military interests in missile production and exports. Moreover, it is unclear how much authority, if any, China's Ministry of Foreign Affairs has over Beijing's arms-export corporations, which are bent on gaining export revenues from all possible sources.[20]

Japan has made a strong government commitment to achieving an independent space capability. While Japan's space program has been dependent largely upon American hardware--in the form of McDonnell Douglas Delta launch vehicle stages--it is steadily advancing toward indigenous capability. The H-II is the first wholly indigenous Japanese SLV, and the Japanese plan to market it for launch services. Japan was among the signatories of the MTCR when it was established in 1987 and has no active ballistic missile development program.

India also possesses operational SLV capabilities. India's SLV-3 booster, patterned after the American Scout sounding rocket, was used to launch the country's first satellite in 1980. An improved version of the SLV-3, the Augmented Satellite Launch Vehicle (ASLV), has been tested three times, with only the most recent a success. India plans to launch the Polar Satellite Launch Vehicle (PSLV), designed to place a one-metric-ton satellite into a 900-km sun-synchronous orbit in 1993.[21] By the end of the decade, India also hopes to test its Geostationary Satellite Launch Vehicle (GSLV), capable of delivering a 1,900-kg payload to geosynchronous orbit, giving New Delhi a potential capability currently enjoyed by only the United States, Russia, China, and the European Space Agency. One impediment to Indian progress, however, may come from international strictures on missile activities. In May 1992, the United States imposed sanctions on the Indian Space Research Organisation because of its contract to purchase cryogenic engine technology from Russia. But it is currently unclear to what degree US sanctions will hinder the Indian program over the long term.

A number of other Asian states, such as South Korea, Indonesia, and Taiwan, have also expressed an interest in pursuing space launch vehicle programs at one time or another. However, such efforts are not believed to be very advanced.[22]

Converting an SLV into an ICBM

Converting an SLV into a long-range ballistic missile involves replacing the SLV's payload with a warhead and reentry vehicle. In order to develop an operational ballistic missile capability, a state must possess a warhead which is both small and light enough to be carried by missile (that is, in the neighborhood of 500 to 1,000 kg). While creating a chemical or biological warhead of such dimensions is not difficult, fielding a compact nuclear weapon is much more so.

In addition, the space launch vehicle needs to be equipped with a reentry vehicle to shield its warhead from atmospheric effects such as heating. A high drag/low-accuracy warhead would not require an advanced heat shield. Reentry vehicle technology is commercially available, although it is controlled by the MTCR. Sounding rockets configured to conduct micro-gravity experiments, for example, possess their own heat shields. In addition, a developing state might be able to construct a fiber/resin heat shield, such as those fielded by the United States and Soviet Union, using commercially-available technology and design information which can be found in open literature. Such a design would add between 75 and 100 pounds to the weight of the warhead and would be capable of protecting nuclear, biological, and chemical payloads.[23]

Most SLVs already possess guidance systems sufficient to allow them to strike a large area target such as a city. A third world ICBM would not necessarily require high accuracy to be useful in military operations. For some objectives, all that may be required would be the ability to deliver missile strikes into an urban area--requiring accuracy on the order of 10 kilometers. Such accuracy is feasible with commercially-available inertial navigation systems, even at intercontinental ranges. In addition, other means of target location, such as data from the Global Positioning System (GPS) satellites, could be used both to locate launchers and to guide missiles.

A force consisting of several liquid-fueled missiles may satisfy the perceived security needs of some states. But states seeking a more sophisticated long-range ballistic missile force would face more extensive challenges and constraints. A larger and survivable missile force might require a shift to solid-fueled missiles and enhanced mobility or hardened launch sites to protect against enemy attack.

In sum, any state capable of fielding an indigenous SLV inherently is able to convert that launcher into a long-range missile fairly rapidly and without enormous effort or expertise. The technology needed is widely available internationally, and the skills required are those which could be routinely acquired in an SLV or satellite development program. A more sophisticated operational missile force, however, would require considerably more investment and technical capability.

Shortfalls of the MTCR

To date, US policies to control proliferation have centered upon restricting the supply of technologies internationally. The cornerstone of US missile nonproliferation efforts remains the 1987 Missile Technology Control Regime, a set of coordinated export policies designed to limit the spread of missiles capable of delivering a 500-kg payload to a distance of 300 km. The MTCR export restrictions include a ban on the sale of missile production facilities and a strong presumption to deny exports of complete delivery systems, including complete rocket systems, such as ballistic missile systems, space launch vehicles and sounding rockets; unmanned air vehicle systems, including cruise missile systems, drones and remotely-piloted vehicles; and the following major subsystems: individual rocket stages; reentry vehicles; rocket engines; guidance sets; thrust vector controls; and warhead safing, arming, fuzing and firing mechanisms (Category I systems). In addition, the export of dual-use missile components (Category II systems) is to be judged on a case-by-case basis.[24]

The MTCR faces a number of limits on its ability to stymie the diffusion of missile technology. First, simply the existence of a cartel limiting the supply of prescribed technologies can raise the economic incentive for cheating by both members and non-members. The Argentine-Egyptian-Iraqi Condor II program, an effort to jointly develop an IRBM which was scrapped in 1990, serves as testimony to the willingness of some European companies to violate even their own nations' export policies. Similarly, US and European firms are known to have supplied equipment that helped Iraq construct its Sa'ad 16 missile development complex, albeit sometimes unwittingly.[25]

Second, restrictions on channels of supply can foster the establishment of new sources of supply operating outside the cartel. With the larger powers moving out of the missile export business, second-tier suppliers such as China and North Korea have sought new market opportunities for their missile systems and components.

A third weakness of the regime is that there remain significant differences in enforcement standards both within and among MTCR members. This flaw is accentuated by the fact that the MTCR is an informal, voluntary association which lacks an institutionalized arrangement to govern interpretation and enforcement of restrictions. For example, the stated goal of the regime is to halt the spread of "nuclear-capable" missiles capable of carrying a 500-kg payload 300 kilometers.[26] While the United States considers any system capable of meeting these technical parameters de facto nuclear-capable, other MTCR adherents, such as France, demand that the state purchasing the missile technology have an active nuclear weapons program for the missile to be considered nuclear-capable. Further, while space-launch vehicle and sounding rocket technology are explicitly covered by restrictions in the regime, and are treated as such by the United States, other MTCR states are not as strict in their interpretation.[27]

A fourth flaw of the regime is its relatively limited membership. The list of MTCR members has expanded significantly in the six years since it was first announced. Currently 22 nations have agreed to join and support the MTCR export guidelines. At a meeting in Canberra in March 1993, Iceland, Hungary and Argentina expressed a desire to join the regime as well.[28] Russia has announced that it will observe the terms of the MTCR and has published its own set of export controls.[29] Although China has agreed to abide by the terms of the MTCR, Beijing reportedly continues to export missile technology.[30] Perhaps more significantly, a tier of states in the developing world which are acquiring an indigenous missile production capability are also unlikely to join an arrangement which they see as pitting the "haves" against the "have-nots".

The debate on missile non-proliferation policy has centered upon mechanisms to modify, supplement, or replace the MTCR. A number of analysts have argued that the MTCR can be effective if modified. In general, they have called for measures to improve its effectiveness by increasing the ranks of its adherents, expanding the list of technologies covered by the regime, or providing the regime with a formal enforcement mechanism.[31]

A more ambitious approach would involve supplementing the MTCR with formal or informal confidence- and security-building measures (CSBMs) between regional rivals in an attempt to reduce the demand for such systems. Among existing measures of this kind, India and Pakistan have negotiated an agreement not to attack one another's nuclear facilities, and Argentina and Brazil have initiated on-site visits to each other's nuclear facilities. Further CSBMs could include information and intelligence exchanges, on-site inspection of defense production and space-launch facilities, and prior notification of missile tests.[32]

A still more ambitious approach would involve supplementing or replacing the MTCR with a formal arms control regime encompassing both missile suppliers in the developed world and the states in the developing world which seek access to such technology. The 1987 U.S.-Soviet INF Treaty has been cited as a model for such arms control efforts.[33] Progress towards such goals is likely to be difficult, however. Such a plan would require a significant degree of cooperation between states with long histories of conflict. As the experience of attempts at creating a nuclear-free zone in the Middle East show, such proposals often become embroiled in local conflicts and power asymmetries. In short, if a state feels a compelling need to acquire ballistic missiles, it will likely abstain from such a regime. In the case of the Non-Proliferation Treaty, for example, those states that wish to acquire nuclear weapons have simply refused to sign the treaty. The recent example set by North Korea demonstrates that even when a state subscribes to an agreement, its continued participation in such a regime cannot be taken for granted.

Furthermore, proliferation increasingly involves non-state actors which are difficult to accommodate within traditional non-proliferation frameworks. The phenomenon of the spread of ballistic missile technology encompasses not only the sale of complete missile systems from suppliers in the developed (and, increasingly, the developing) world, but also through the sale of missile components by firms in a variety of states. The MTCR deals with this problem by making it the responsibility of the exporting state to obtain binding assurances as to end-use of the export from the recipient state.[34] However, there have already been several instances of private concerns evading these controls. The most significant of these has been the Consen group, which has operated out of Monaco. The group coordinated the Condor II program to build a 1,000-km missile for Argentina, Egypt and Iraq through a network of companies throughout Europe.[35]

In addition, states may not behave as unitary actors in all cases. One example is the difficulty Argentine President Carlos Menem has had in halting his country's involvement in the Condor II project: Although he publicly renounced the program, begun under his predecessor, the Argentine Air Force reportedly obstructed attempts to dismantle the infrastructure used to manufacture the Condor and to destroy the missiles which were produced.[36]

A final impediment is the difficulty of controlling the flow of individuals and their knowledge. It has been widely reported that Iraq modified its Scud missiles with the help of foreign missile experts.[37] Similarly, Iraq employed the services of Canadian artillery expert Gerald Bull to build the "supergun." Brigadier General Hugo Olivares Piva, former director of the Brazilian Aerospace Technology Center (CTA), heads a group of rocket scientists which helps foreign countries, including Iran and Iraq, with missile programs.[38] Most ominous is the prospect of scientists and engineers from the former Soviet Union assisting third world states in developing ballistic missiles. For example, the Russian press recently reported that a group of defense workers "engaged in problems of strategic missile armaments" were prevented from emigrating to North Korea because they planned on assisting Pyongyang's arms efforts.[39]

Conclusions

The relative weaknesses of the MTCR and the limited utility of traditional arms control measures for controlling the diffusion of missile technology suggest that the pursuit of a missile restraint regime should be subsumed into ongoing efforts to end regional conflicts; and, in the interim, should focus on more modest instruments aimed at confidence-building. Confidence- and security-building measures, including information and intelligence exchanges, on-site visits of defense production and space launch mechanisms which promote consultation among regional rivals could help ease unwarranted suspicions about missile production efforts, limit their political and military consequences, and, possibly, reduce some of the incentives now propelling the expansion of these problems. CSBMs can reduce tensions by mitigating the mystery about rivals' military activities, providing channels for routine interaction, and demonstrating adversaries' interest in reassuring other states about their military objectives. Although these instruments are only valuable as indicators of political will and can be violated at any time, they can serve as the beginnings of a diplomatic infrastructure needed for broader accommodation.

Declaration of intent, like a pledge not to use ballistic missiles preemptively, obviously might not endure in a crisis, but they are nevertheless signs of political conciliation which should not be dismissed out of hand. Similarly, on-site visits and prior notification of test launches do nothing to stop dedicated missile programs, but they can help to reduce the climate of suspicion among adversaries through increased communication.

The United States, for example, has been exploring these kinds of initiatives with Middle East partners for several years. In late 1988, the United States reportedly held discussions with both Egypt and Israel concerning missile-related confidence-building measures. Among the proposals mentioned were notifications of missile launches, whether planned missile tests or practice firings of operational systems. Some U.S. government officials also advocated that countries in the region adopt a "no first use" policy. Some of these measures may be appropriate for other regions such as South Asia and Northeast Asia. Other CSBMs which could be considered for missile restraint include application of international safeguards and on-site verification at space launch facilities to ensure they are not being used to develop missiles; regional export controls, such as agreements not to sell missiles to unstable states; and routine bilateral military exchanges between rival states to discuss common security concerns.

Additional measures should address the role nominally civilian space programs have played in the development of ballistic missile capabilities. In particular, policy makers should examine mechanisms to allow developing states access to space without providing them the means to develop long-range missiles. One such measure would be the development of an international space launch agency to give countries access to space in return for a pledge not to produce their own space-launch vehicles. Such an arrangement would likely be appealing to developing states, especially if space launches are offered at attractive rates.

Agreements to delimit missile deployment areas--moving forces away from borders, for example, and declaring fixed deployment sites which could be subject to monitoring--could reduce the perceived threat of surprise attack and, in principle, could be a stabilizing measure. Deployment limitations could be an important element of a more comprehensive regional or bilateral security pact. The Sinai Agreement, which provides for peacekeeping forces and other enforcement mechanisms to monitor proscribed military activities in the area, is an important achievement which has enhanced stability and mitigated tensions between Israel and Egypt since 1974.[40]

As noted already, achieving more significant curbs on the demand for missiles will depend on progress in the reduction of overall regional tensions. This broader objective can be fostered by encouraging states to pursue incremental measures aimed at enhancing confidence. The United States can play an important role in encouraging regional powers to pursue CSBMs, although the choice of initiatives must ultimately come from the states themselves.

As a first step, the United States can take the lead in helping countries to develop routine consultative mechanisms for exchanges of information about military programs, discussions of mutual security concerns, and, over time, consideration of more ambitious arms control measures. The United States is a source of leadership and operational expertise about such mechanisms which are often genuinely unfamiliar to developing states.

Indeed, even US assistance in such prosaic areas as customs enforcement, automated data collection for assessing force balances, or mechanisms to monitor exports, can prove useful. The effectiveness of US diplomatic efforts will require avoiding an exaggerated political profile, almost always the undoing of sensitive diplomacy. Moreover, the United States and its allies should not bear disproportionate responsibility for encouraging restraint. Other missile suppliers, including the former Soviet Union, PRC, and neutral states, must be induced to cooperate.

A complementary approach which might control missile proliferation would be to reduce their military utility with defenses. The use of the Patriot theater missile defense system to protect both Saudi Arabia and Israel during the Gulf War was at least a partial demonstration of the potential of antimissile defenses, albeit still controversial. The Patriot served a dual role: Militarily, it helped protect coalition forces and their host countries from attack by Iraqi missiles; politically, it provided reassurance to countries facing Iraqi attempts at coercion. One approach to building regional confidence might be to encourage states to agree to trade ballistic missiles for theater missile defenses. Such an approach is unlikely to find support among potential aggressors, but stands a chance of being welcomed by countries which may become objects of aggression. Trading offensive weapons for defensive systems may not prevent a state from launching an attack on its neighbors, but could reduce incentives for missile acquisition if implemented by mutual and verifiable agreement.

For all of its potential to help countries defend against aggression, there is nevertheless a danger in overselling TMD as the solution to the risks posed by proliferation. The sale of defensive systems and technology to states which have or are trying to develop missile production capabilities could indirectly contribute to proliferation by granting these countries access to technologies and expertise useful for developing offensive systems. These range from guidance and rocket components to testing equipment and expertise about the phenomenology of missiles.

South Korea, for example, succeeded in modifying the US Nike-Hercules air defense system into a ballistic missile, a program which it pursued despite strenuous U.S. objections. As has been recognized in the US-Israeli Arrow TMD program, the risk of misapplication of defensive technology is sufficiently high to warrant careful controls of such programs. End use controls will have to be applied in future decisions allowing the transfer of advanced defenses as well.

In addition, spreading theater missile defenses among military rivals could prompt them to augment their offenses to overcome defenses deployed by their adversaries. In particular, they may seek to develop countermeasures to theater missile defenses. The propensity of the Iraqi Al-Hussein missile to break up upon reentering the atmosphere posed a challenge to Patriot. In effect, the missile's performance inadvertently acted like that of decoys or chaff, confusing and potentially overwhelming the defense. Even if Patriot was not designed originally to intercept ballistic missiles, its limited effectiveness against 1960s-vintage missiles operated by a technically unskilled military suggests that increasing the effectiveness of defensive systems against countermeasures will be a continuing challenge.[41]

Nonetheless, ballistic missiles have enjoyed a privileged position as the only weapon against which there is no deployed defense. The fielding of active defenses, if effective, might substantially reduce the attractiveness of the ballistic missile as both a weapon and an instrument of political coercion. TMD alone can hardly be expected to solve the problems posed by the proliferation of weapons of mass destruction and their means of delivery. However, the deployment of defenses could help discourage what is currently an easy option for states seeking a coercive capability and, in turn, help channel third-world arms purchases down less destabilizing paths.

In order for arms control to become attractive to countries in various regions, it must be demonstrated that such agreements offer benefits which are not attainable through other means. Most countries in South Asia or the Middle East have limited familiarity with arms control concepts and are suspicious of negotiated security arrangements which require reductions in military capabilities. The means by which third-world countries might be encouraged towards non-aggressive postures as they acquire new weapon capabilities are not well understood. The predominant focus of industrial countries' policies for containing regional missile arsenals has been to prevent technology from proliferating, not on what to do once prevention has failed. But efforts to restrain missile programs are most likely to be effective if pursued as part of initiatives to end or contain regional conflicts, one of several instruments intended to manage the transition to a more codified approach to resolving disputes peacefully.

Part of the enduring challenge of controlling proliferation also stems from the absence of any agreed, workable definitions of the security threats posed by the spread of missiles or other advanced delivery capabilities. Although it is largely undisputed that the development of nuclear capabilities around the world should be controlled, and the use of chemical and biological weapons even on the battlefield is despicable, the relative legitimacy of conventional arms sales--including missiles--remains a matter of controversy.

Conventional weapons have always been seen as the benign alternative to nuclear proliferation and serve as a common instrument of dissuasion in efforts to stop new states from going the nuclear or chemical route. Other than the Missile Technology Control Regime, there is no formal international apparatus to guide transfers of conventional technologies to the Third World. Despite their pertinence for the delivery of nuclear and chemical weapons, governments have resisted placing controls on transfers of combat aircraft and on most dual-use technologies. With the exception of the list of targeted technologies contained in the MTCR technology annexes, weapons production technology is continuing to spread without benefit of formal international review, let alone coordination.

Another challenge to effective non-proliferation arises from the growing percentage of advanced technology needed for military products and innovations emerging from the commercial sector. Leading-edge technologies, from fiber optics to microcircuitry to advanced software, increasingly are produced by commercial enterprises which are not necessarily accountable to governments. For this reason alone, it is clear that no central regime can survive if it is perceived as excessively penalizing to private enterprises. However lofty the goals, any policy which appears unduly injurious to economic competitiveness cannot endure. Devising criteria for dual-use exports to the Third World, in particular, will require difficult choices about desirable and undesirable types of proliferation, disaggregating technologies which are useful for development activities from their military applications.

Eliciting the support of industry will be a vital element of the success of an enduring military technology export regime. This, in turn, will require that controls be multinationally supported and highly selective. Industry can play a key role in helping to develop the lists of items and technologies that are to be controlled, to compile information about sources of technology, and to design and implement workable security safeguards which do not interfere with desirable private enterprise.

The role of the American Chemical Manufacturers Association in the chemical weapon convention negotiations in Geneva may be an apt model for other areas of technology transfers. The chemical industry has been serving as a vital source of expertise for negotiators, identifying technologies and inputs to include in the treaty, and helping to devise practical verification schemes. It is obviously in their self-interest to influence the scope of agreed controls, and to be perceived as supportive of a chemical weapons ban.

Similarly, computer and civilian space companies could stand to lose the most from any draconian measures imposed as a result of heightened international concerns about the diversion of these kinds of technologies for missile development or other offensive military uses. It would be in the immediate self-interest of such companies to assist governments to restrain missile programs in problematic states by helping to identify relevant technological inputs needed for missile development and in devising safeguards which can discourage the adaptation of civilian equipment for military programs.

As the main source of expertise about technology and usually the party most involved in actual transactions, industry may be the only means by which governments can track and enforce restrictions on exported products. By the same token, the perception that industry is not cooperating in non-proliferation efforts could impose penalties exceeding the revenues foregone by refusing certain foreign contracts.

But in the final analysis, the main reason that the global proliferation problem is as serious as it is today is that governments have not pursued non-proliferation policies vigorously. Concerns about proliferation typically have been held hostage to other foreign policy priorities. In countless cases, from Pakistan to Iraq, the West has looked the other way when proliferation occurred because other diplomatic objectives were deemed more pressing. For all of the obvious costs of this policy revealed after Iraq, this attitude apparently still prevails among industrial countries with respect to important allies. Absent cooperation and among the major suppliers to invest effort in the enforcement of controls and the willingness of key recipients to accept these norms, international restrictions on missile or any other proscribed technologies will obviously not endure.

The proliferation of long-range ballistic missiles is not inevitable. By understanding the motivations that may lead other states to develop, deploy, and employ ballistic missiles, we may be better able to prevent other nations from acquiring them. Measures taken today to increase the costs and decrease the benefits of acquiring ballistic missiles will reduce the threat to the United States, its forces, friends and allies in the future.

Notes

1. The views expressed in this paper are those of the authors and do not reflect the official policy or positions of the Department of Defense or the U.S. Government.

2. B. Chow, "Third-World Space Launch Programs: Economics and Safeguards," RAND Report R-4179-USDP, January 1992.

3. The views expressed in this paper are those of the authors and do not reflect the official policy or positions of the Department of Defense or the U.S. Government.

3. B. Chow, "Third-World Space Launch Programs: Economics and Safeguards," RAND Report R-4179-USDP, January 1992.

3. K. Subrahmanyam, "The Meaning of Agni," The Hindustan Times (New Delhi) in English, June 2, 1989.

4. "Resource Satellite to be Built with PRC," Rio de Janeiro O Globo in Portuguese, December 21, 1989, p. 30.

5. J.R. Harvey, "Regional Ballistic Missiles and Advanced Attack Aircraft: Comparing Military Effectiveness," *International Security,* vol. 17, no. 2 (Fall 1992), pp. 41-83.

6. Statement of Mr. Henry D. Sokolski, Deputy for Non-Proliferation Policy, Office of the Assistant Secretary of Defense, International Security Affairs, Department of Defense, before the Joint Economic Committee,

Subcommittee on Technology and National Security, "The Military Implications of Proliferation in the Middle East," March 13, 1992, p. 4.

7 Speech by Col. Khaddafi at a meeting of students of the Higher Institute for Applied Social Studies at the Greater al-Fatih University, April 18, 1990, Tripoli Television Service, April 19, 1990

8. "Preparing for Protection of our National Interests," *Resalat,* December 31, 1990; "A Military Lesson from the Persian Gulf War," *Kayhan International,* March 17, 1991.

9. Air Commodore J. Singh, "Let Us Not Be Shy of Agni," New Delhi, *The Hindustan Times* in English, April 19, 1989. See also R. Singh, "Advanced Weaponry for the Third World" in E.H. Arnett, ed., *Science and International Security: Responding to a Changing World* (Washington: AAAS, 1990), p. 8.

10. J. Lewis and Hua Di, "China's Ballistic Missile Programs: Technologies, Strategies, Goals," *International Security,* vol. 17, no. 2 (Fall 1992), pp. 34-35.

11. "Defense Ministry Cites DPRK Missile Upgrades," Seoul, *The Korea Herald* in English, September 9, 1992, p. 3.

12. J.S. Bermudez, Jr., "Syria's Acquisition of North Korean 'Scuds'," *Jane's Intelligence Review,* vol. 4, no. 4 (April 1992).

13. J.S. Bermudez and W. Seth Carus, "The North Korean 'Scud B' Programme,"; J.S. Bermudez, Jr., "Ballistic Missiles in the Third World-- Iran's Medium-Range Missiles," *Jane's Intelligence Review,* vol. 4, no. 4 (April 1992).

14. Lewis and Di, "China's Ballistic Missile Programs," p. 37.

15. W. Seth Carus, "Long-Range Rocket Artillery in the Third World," *Jane's Intelligence Review,* vol. 3, no. 10 (October 1991), p. 476.

16. See S. Graybeal and P. McFate, "GPALS and Foreign Space Launch Vehicle Capabilities" (McLean, VA: Science Applications International Corporation, February 1992); A. Karp, "Ballistic Missiles in the Third World," *International Security* (Winter 1984-85); T.G. Mahnken, "Why Third World Space Systems Matter," *Orbis,* vol. 35, no. 4 (Fall 1991).

17. T.G. Mahnken and T.D. Hoyt, "The Spread of Missile Technology to the Third World," *Comparative Strategy,* vol. 9, no. 3 (Fall 1990), pp. 245-263.

18. *Decisions Maker's Guide,* pp. 21, 141; Government of India, Department of Space, *Annual Report 1989-90,* p. 34; "U.S. Imposes Sanctions Against Russian, Indian Concerns Over Rocket Deal," *The Washington Post,* May 12, 1992, p. A15.

19. *Interavia Space Directory, 1990-1991* (Alexandria, VA: Jane's Information Group, 1990).

20. J. Lewis, Hua Di, and Xue Litai, "Beijing's Defense Establishment," *International Security,* vol. 15, no. 4 (Spring 1991).

21. "U.S. Ban Could Delay Some Satellite Launches," Madras, *The Hindu* in English, June 18, 1992, p. 9.

22. Mahnken and Hoyt, "The Spread of Missile Technology to the Third World," p. 573; *Decisions Maker's Guide to International Space,* (Arlington, VA: ANSER, 1992), pp. 165-167.

23. Graybeal and McFate, "GPALS and Foreign Space Launch Capabilities," pp. 7-8.

24. White House Office of the Press Secretary, "Missile Technology Control Regime Guidelines for Sensitive Missile-Relevant Transfers," April 16, 1987.

25. "U.S. Firms Helped Iraq Gain Ability to Make Missiles, Officials Say," *The Washington Post,* May 3, 1989, p. 19.

26. White House, Missile Technology Control Regime Guidelines.

27. K.C. Bailey, "Arms Control for the Middle East," *International Defense Review* (April 1991), p. 10; Marc S. Palevitz, "Beyond Deterrence: What the US Should Do About Ballistic Missiles in the Third World," *Strategic Review,* vol. 18, no. 3 (Summer 1990), p. 52.

28. "Move Against N. Korea Seen as Turning Point in Arms Control," *The Washington Post,* April 7, 1993, p. A23.

29. JPRS Proliferation Issues, "Russian Federation Missile Export Control Legislation" JPRS-TND-93-0002-L, March 3, 1993.

30. "China Breaking Missile Pledge, U.S. Aides Say," *The New York Times,* May 6, 1993, p. A1.

31. M. Navias, "Ballistic Missile Proliferation in the Third World," *Adelphi Papers* no. 252 (Summer 1990), pp. 64-68; Janne E. Nolan, *Trappings of Power: Ballistic Missiles in the Third World* (Washington: Brookings, 1991), pp. 145-55.

32. Nolan, Janne E., "Ballistic Missiles in the Third World--The Limits of Nonproliferation," *Arms Control Today,* vol. 19, no. 9 (November 1989) pp. 13-14.

33. Bailey, "Can Missile Proliferation Be Reversed?," pp. 311-314; M.M. Kampelman and E.C. Luck, "Ban Missiles in the Middle East," *The Washington Post,* April 18, 1991, p. A21; M. Nacht, J. Winik, and A. Platt, "What About Arms Control?" *The Washington Post,* September 22, 1991, p. C3.

34. White House, Missile Technology Control Regime Guidelines.

35. "The Flight of the Condor," London, *The Financial Times* in English, November 21, 1989, p. 1.

36. "Argentina's President Battles His Own Air Force on Missile," *The New York Times,* May 13, 1991, p. 1.

37. "Engineers Comments on Iraqi Technology Capability," Sao Paulo, *Folha de Sao Paulo* in Portuguese, February 6, 1991, p. A12.

38. "Missile Experts Want to Return from Iraq," Sao Paulo, *O Estado de Sao Paulo* in Portuguese, September 2, 1990, p. 18; "Retired Officer Proposes Missile Deal with Iran," Sao Paulo, *Folha de Sao Paulo* in Portuguese, February 22, 1991, p. 4.

39. "Specialists Prevented from Leaving DPRK, " Moscow ITAR-TASS World Service in Russian, 1620 GMT, February 9, 1993.

40. The limitations on weapon deployments, however vital, are only one facet of a broader political accommodation, made possible only after Egypt and Israel achieved a political understanding in the Camp David accords. For now, no other Middle Eastern military rivals seem promising cases for such arrangements.

41. "U.S. Reviews Missile Plan," *Defense News,* August 12, 1991, p. 1.

2

Space Power and Space Interests: United States

John Pike
and Eric Stambler

Outer space has unique psychological significance. All nations and peoples regard the glory of the same celestial firmament. The space environment imposes unique technological challenges to those who would attain and utilize the space environment. No place on Earth is so difficult or costly to reach, and the technical challenges and financial burdens of going into space far exceed those of even the most inaccessible terrestrial environment. The ocean depths and the high mountains, the frigid poles and torrid deserts, pose trivial challenges compared to those of space. The requirements of space operations test even the most technically advanced and wealthiest societies, and far exceed the grasp of most of Earth's nations.

Thus, the states of the world are divided into at least three classes of space capabilities. Only the United States and Russia possess the full range of small and large launch vehicles, piloted and robotic spacecraft, and military and civil space capabilities. A growing number of other states possess some but not all of these capabilities, typically consisting of smaller launch vehicles capable of orbiting robotic spacecraft for scientific and other largely civilian applications. The vast majority of states are not space powers, and derive the benefits from the exploitation of space only through the capabilities of others.

The club of space powers has steadily expanded over the past 35 years, and there is every reason to expect further expansion in coming decades. Space, once the province of the superpowers, has come within the reach of states with diverse levels of economic and technological capability and achievement.

That we are "all under heaven" provides a powerful image of our common humanity, of our common destiny on Earth, and in space. This fundamental nature of the space environment gives impetus for global application of space technology. But the majesty of the heavens also endows weapons in space, or those that pass through the space environment (such as ballistic missiles) with the capability to cause a unique psychological dread.

Security Concerns

At least four security concerns have been raised by various states about the potential directions of military space activities. These concerns are interrelated, which complicates their resolution.

Diffusion of Rocket Technology

One concern derives from the use of rocket technology for the delivery of nuclear weapons and weapons of mass destruction. The military was among the earliest users of rocket technology, and some of the greatest advances in rocketry were achieved in times of war. The first triumphs of the space age were accomplished using rockets developed for long-range delivery of nuclear weapons, and there continues to be a close relationship between rockets used for weapons delivery and rockets used for space applications. The rocket forces of the United States and Soviet Union were among the more visible manifestations of the Cold War. Negotiations to limit these forces were a central element of Soviet-American relations for over three decades, and further reductions in these forces continue to have high priority. Given this history, it is not surprising that the growing number of rocket-capable countries should be a focus of attention.

Anti-missile Systems

Another concern is a potential response to the long range missile threat --the development and deployment of anti-missile systems, including anti-missile weapons based in space. The weaponization of space, and the potential for an arms race in outer space, have been of growing concern to the international community in recent years. Research on the technologies needed to intercept long-range missiles began simultaneously with the development of long-range missiles, in the years following the Second World War. For over three decades the United States and the Soviet Union actively worked on the development of anti-missile technologies, but did not deploy large-scale anti-missile systems. During the Cold War, it was widely perceived that deploying anti-missile systems could be destabilizing--that such systems would provoke further deployments of offensive missiles, and provoke a preemptive attack in a time of crisis. Thus the deployment of large-scale anti-missile systems was widely seen as risking the extension of the arms race into outer space, its continuation on Earth, with all the associated risks to both participants and observers. The emergence of new anti-missile technologies, the evolving relationship between the United States and Russia, and the increasing number of countries that may deploy long-range missiles in the future have suggested to some that these concerns may no longer be relevant. To others, it seems evident that the prevention of an arms race in space may no longer preclude the deployment of anti-missile weapons on the Earth 'or in space, if this can be accomplished in a cooperative manner without precipitating an open ended arms competition. However, these new perspectives remain controversial.

Space Support for Ground Operations

A third concern is related to the increasing application of military space systems to support terrestrial combat operations, and the significant disparities in such capabilities. In the early years of the space age, military space systems were largely used for peacetime support functions, pre-war planning and treaty verification. However, with the improved capabilities of more modern systems, military satellites are increasingly relevant to providing direct support to the contemporary battlefield. While the full extent of the contribution of American military satellites to the outcome of Operation Desert Storm is still being debated, it is clear that the greatest disparity in military capabilities in this conflict was in the field of space systems. In the past, Americans worried about the combat capabilities of Soviet military satellites. Today, other countries express similar concerns about the preponderance of American military space systems. In the future, the United States, along with other states, may also view with apprehension the military space capabilities of rising space powers.

Anti-satellite Weapons

A fourth potential pathway to militarizing space is the possible deployment of anti-satellite weapons. The growing relevance of space to terrestrial combat operations challenges the assumption that space should remain a sanctuary in which satellites are able to operate free from the risk of attack. When space systems were little more than peacetime adjuncts, there were perhaps few incentives to attack them. But as military satellites have become vital to the success of terrestrial combat operations, military planners face increasing incentives to negate these elements. Unfortunately, anti-satellite weapons share many of the characteristics of anti-missile weapons, and the deployment of one may facilitate the deployment of the other. Anti-satellite weapons could also unleash a destabilizing arms race, as well as increasing the risk of preemptive attacks in a time of crisis. These concerns are magnified by the fact that modern military satellites are generally few in number and very expensive. Thus, a modest force of anti-satellite weapons, acquired for a fraction of the cost of the satellites they would attack, could have a disproportionate impact on a developing military situation. While anti-satellite systems have remained the province of America and Russia, emerging space powers may also consider developing anti-satellite weapons.

These concerns must also be balanced by the significant opportunities in space. The exploration and use of space has been characterized to date by an unusually high degree of international cooperation. The process of reducing reliance on the use of force for the settlement of international disputes, of reducing the level of armaments, and for improving cooperative mechanisms for conflict management, has become more and more important in the closing years of the 20th century. There are now new opportunities to fulfill the United Nations Charter, "to save succeeding generations from the scourge of war... which has brought untold sorrow and suffering to mankind." Outer space, which has largely escaped this scourge, may help to resolve conflicts in the 20th century and usher in a new era of cooperation in the 21st century.

A New Regime

Two converging developments suggest the need for new international agreement on access to space launch technology. The first is the proliferation of long-range missile technology to additional countries, frequently under the guise of civilian space programs. The second is the entrance into the international market for space launch services by countries with non-market economies such as Russia and China. A common response to these twin challenges might be patterned on the International Atomic Energy Agency, which is simultaneously responsible for controlling the military applications of nuclear technology, while promoting its civilian use. A "Rockets for Peace" program might follow in the steps of President Eisenhower's "Atoms for Peace" program.

Rockets for Peace

This new international regime would consist of four elements: (1) agreement by all parties not to transfer missile technology to non-spacefaring nations, along the lines of the existing Missile Technology Control Regime; (2) a commitment by non-spacefaring nations not to develop or test rockets with capabilities in excess of those proscribed by the MTCR; (3) a commitment by the spacefaring nations to a common pricing structure under which they will sell launch services, similar to existing marketing agreements covering air and marine transport; and (4) a commitment to donate launch services to less-developed countries.

This approach directly addresses the domestic political base for missile programs in many third world countries, by severing the link with civilian space development. The recent transfer by Argentina of responsibility for the Condor project from the military to scientific authorities demonstrates the importance of this approach.

This regime would meet simultaneously many of the concerns of all parties. The West would obtain Russian and Chinese adherence to the Missile Technology Control Regime, as well as agreement by these countries to space launch marketing policies that would not undercut Western launch vehicle marketing. Russia and China would gain greater and more regular access to the international market for launch services. And less developed countries would gain assured access to space without the expense of having to develop indigenous launch capabilities.

Russian Stability

The most immediate application of this "Rockets for Peace" approach is found in the Clinton Administration's decision in 1993 to consolidate the Russian and American space station efforts. With the decision to join the Freedom and Mir space stations, Bill Clinton has restored a purpose to America's space program. In 1961, John Kennedy initiated Apollo, as part of the Cold War competition with the Soviet Union. Three decades later, Bill Clinton is using space cooperation to define a new relationship with Russia. From the twin perspectives of national security policy and space policy, this policy is an historic achievement.

It is urgent to convert the former Soviet aerospace complex due to the potential proliferation threat--both domestic and international--posed by the personnel of the complex. Domestically, unemployed aerospace workers are potential supporters for those who seek to reverse the recent reforms, and restore the authoritarian and militaristic old regime. Internationally, migrating aerospace engineers increase the risk of proliferation of missile technology to hostile regimes. Although Russia has agreed to adhere to MTCR guidelines, the other republics have not. This threat of missile proliferation (which has not received significant attention in the West) is greater than that of nuclear proliferation (which has received considerable attention), given the relatively simple industrial infrastructure needed to support a missile program. A nuclear weapons program requires fairly elaborate and sophisticated support facilities, while missiles require little more than standard machine shop tooling.

In the 1970s American aerospace companies failed dismally in their attempts to convert to civilian products such as buses and subway cars. More recent diversification through acquisitions have been equally disappointing. The experience in the former Soviet Union has been equally discouraging. Thus, in the absence of other outlets, it is likely that former Soviet aerospace enterprises will remain a focus of advocacy for a return to a more hostile posture against the West.

The former Soviet aerospace industry has a substantial potential to compete successfully in the global market for military and civilian aircraft, as well as space systems and services. But sales of arms to the Third World will only further exacerbate regional tensions. And sales of civil aircraft and space products will compete directly with sales by American companies. There is no doubt that in the long run the aerospace complex of the former Soviet Union will find a place in the world market. But in the near term, international sales are likely to remain both too meager to affect the course of events at home, while simultaneously looming quite large to American companies affected by international competition.

It is essential to harness the talents and energies of the former Russian aerospace industry to the cause of international cooperation. Joint construction of a space station would at a stroke reduce the sources of the aerospace industry's support for renewed international tensions, and condition their future prosperity on further reform. If this effort is successful, it will demonstrate the potential for space cooperation to all nations.

This new partnership in space began in January 1994 with the first flight of a Russian cosmonaut on the Space Shuttle--the prelude to as many as ten American Shuttle missions to the Mir space station through 1997. Construction of the new international space station will begin with an initial launch of a Russian Mir core module in late 1997, followed by the launch of elements derived from Freedom. The station would be completed in 2002, with a permanent complement of six crew members.

The crucial issue is not the building of the station, however. Rather it is how to engage the Russian aerospace complex in a cooperative venture that will have benefits beyond the reward of joint space exploration. The consolidation of the Russian and American piloted spaceflight efforts represents a unique and highly visible exemplar of the new partnership between these former adversaries. During the Cold War, the space race represented a continuing reminder of the bipolar competition. Space achievements epitomized national aspirations and identities in both countries. There is no more effective vehicle for demonstrating the fundamental change in the relationship between America and Russia than cooperation in human space flight.

American Space Interests

American interests in space include the application of military space systems to support terrestrial military operations; the use of space weapons to counter ballistic missiles; and commercial uses of space, such as launch vehicles. The diversity of these interests greatly complicates the formulation of a coherent policy, but the interrelationship of these fields provides both the opportunity and necessity of an integrated strategy.

Military Space Systems

The United States entered the Gulf War with an unprecedented number of operational space systems in orbit. If the war had taken place even three years earlier, it is unlikely that the contribution of military space systems would have been nearly so great. Today, the United States maintains an array of military space systems with capabilities that surpass those of Russia, and dwarfs those of other space powers. From the perspective of the 21st century, the use of American military space systems in Operation Desert Storm may be seen as marking a major watershed in the history of military technology and military tactics, ranking with the introduction of effective armored operations in World War II or the machine gun at the beginning of the century.

The full significance of the contribution of American military space systems to the outcome of Operation Desert Storm cannot be determined definitively. But it is clear that the disparity in military space capabilities was one of the distinguishing features of that conflict. Desert Storm was the first "Space War," since it was the first time that the full range of modern military space assets was applied to a terrestrial conflict. It is equally clear that proponents of military space systems will point to the outcome of Desert Storm as a sign of the decisive potential of military space systems.

An alternative view contends that military space systems were of marginal relevance to the outcome of Desert Storm. Perhaps Desert Storm was not so much a case of a coalition victory through superior technology, as an Iraqi loss due to political and military incompetence. The relative ineffectiveness of the campaign against Iraqi Scud launchers was a manifestation of the limited utility of space systems. During the later phase of the air campaign, target acquisition did not rely on sophisticated satellite systems, but rather on the initiative of individual pilots. And by the commencement of the ground campaign, space systems were largely irrelevant to the conduct of the war.

Which of these contending interpretations is correct? Overall, it is difficult to judge because it is much easier to identify the technological inputs of military space systems in the Gulf conflict than it is to identify the military outputs resulting from their application. It is difficult to say precisely what operational military difference was made by the presence of space capabilities, particularly since in most cases there were other non-space systems providing similar or complementary inputs.

Despite these ambiguities, the primary lesson learned from the Gulf War is that no future war on Earth will be fought without such support systems. From this perspective, the real question is not so much what military space systems accomplished in Desert Storm, but rather what Desert Storm suggested could be accomplished by such systems in some future conflict. Desert Storm may be viewed as analogous to the First World War, which witnessed the baptism under fire of weapons such as the tank and aircraft, although the full combat potential of these weapons was not achieved until the Second World War.

Denying Access to C31

Autonomous space launch capabilities also raise questions about the uses to which satellites may be put, such as military reconnaissance. Permitting the sale of sophisticated satellite technology to states such as the United Arab Emirates would set a dangerous precedent that could justify future sales by other nations to less desirable nations. The possession of such a satellite by the Emirates, for example, could enable other nations to acquire intelligence information from it that runs counter to US interests. For this reason, any efforts by the Commerce Department to move satellites from the Munitions List to allow commercial sales should be scrutinized closely.

Preventing the proliferation of this technology alleviates the need for methods to counter its use. It is a crucial issue whether the sale of surveillance satellites is evaluated in terms of national security concerns, with little or no regard to economic considerations. More security can be gained by diplomatic initiatives to limit the spread of data from space systems than by promoting exports of such systems.

The motivations for third world countries to obtain reconnaissance satellites are unclear. Although the acquisition of a modest capability may cost no more than airborne early warning aircraft sought by many developing nations, the function of these aircraft is for protection from aerial attack. In most cases, the strategic depth of third world nations is shallow enough that the use of satellites would be excessive. Based on the experience of US forces in Desert Storm, any attempts by developing nations to use satellite data for direct combat support are unlikely to be successful. In addition, while airborne early warning aircraft are generally viewed as performing a defensive role, reconnaissance satellites may be used to perform long range targeting far within an adversary's borders.

In the case of most regional rivalries, only very limited long-range strike capabilities presently exist. Most intelligence gathering can be done by airplane and target acquisition can be performed using less sophisticated means. In such circumstances, the introduction of highly capable space systems might be viewed as highly provocative.

The only circumstance in which the use of reconnaissance satellites could unambiguously add to crisis stability would be under symmetrical conditions, as eventually developed between the United States and the Soviet Union. However, recent military history appears devoid of instances where the possession of such systems would have prevented conflict. The occurrence of "bolt from the blue" attacks, which might be better detected by space platforms than by other means, are difficult to identify. In most cases of strategic surprise, intelligence failures resulted from improper assessment of an opponent's intentions, rather than an inability to assess military capabilities.

Asymmetries in intelligence collection capabilities can be exploited. The abiding interest in certain American military circles to develop a robust space control capability springs from these considerations. Such an advantage was a valuable asset in the Gulf War, enabling the United States to persuade Saudi Arabia of the need for a US military presence by providing pictures of Iraqi troop positions. It is unlikely that these advantages have escaped the notice of the developing world. The United Arab Emirates certainly would find itself in a position of increased prestige within the Arab League were it the only nation in possession of such capabilities.

Third World Prestige

American concerns about missile proliferation date back to the mid-1950s, with the emergence of Soviet long-range rocketry. Although the diffusion of sounding rocket technology to the Third World for peaceful space exploration initially proceeded under the auspices of US aid programs in the 1960s, in time the spread of this technology was increasingly viewed with trepidation by American decision-makers. Such technology might assist third world countries in building ballistic missiles. Indeed, Indian and Brazilian programs got their start through such assistance.

Missile proliferation became a renewed concern for the international community during the Reagan Administration. The 1987 Missile Technology Control Regime (MTCR), was an informal agreement among industrialized countries to control the spread of missiles. However, events in the latter half of the decade created the public impression of a problem out of control. In 1988, the Chinese sale of ballistic missiles to Saudi Arabia caught the United States by surprise. Periodic press reports of possible new missile sales, the leakage of technology from the industrialized world and third world development programs added to the impression. Finally, missile attacks on civilian populations during the Iran-Iraq war and by Iraq during the Gulf War brought home the human dimension of proliferation.

The proliferation of nuclear and chemical weapons, as well as the proliferation of ballistic missiles, have emerged as a major insecurity of the 1990s. Although these missiles are unlikely to threaten the American homeland for the foreseeable future, they may pose a more direct threat to American overseas interests and allies.

Some US friends and allies, notably Israel, are properly alarmed by these developments. Even a few rockets armed with chemical warheads could temporarily ground the Israeli air force or terrorize civilian populations, with potentially catastrophic consequences. Similar concerns have been expressed about the potential vulnerability of Europe, Japan and Korea.

By the end of the century over a dozen countries (including Argentina, Brazil, Egypt, India, Iran, Iraq, Israel, Pakistan, Saudi Arabia, South Africa, South Korea, Syria and Taiwan) may have the technological potential to field medium or intermediate range ballistic missiles. Some of these nations also have chemical and nuclear weapons development programs in various stages of advancement. And a few of these states also conduct military operations against their neighbors with distressing regularity.

Some have suggested that the delivery of chemical or nuclear warheads by rockets, coupled with an absence of warning and the high probability of mission success, may prove significantly more destabilizing than the delivery of these munitions by aircraft. However, when considering the dangers posed by these longer-range weapons, it is important not to lose sight of the fact that many third world missile programs (notably Egypt, Iraq and India) are national prestige projects, rather than military weapons. An early Egyptian effort to acquire longer-range rockets in the early 1960s appears in retrospect to have been little more than an attempt to emulate the Soviet practice of displaying rockets in military parades.

This acquisition of prestige weapons by the Third World did not begin with the missile age. Rather it has been a distinguishing characteristic of the 20th century military scene. Lesser regional powers have consistently sought to acquire a token force of whatever passed as the prestige weapon of the day among the great powers. In the years preceding World War I, Argentina, Brazil and Chile enacted a miniature replica of the naval arms race that erupted between Germany and the United Kingdom in 1904. The use of missile projects to enhance national prestige has ample precedent in Soviet behavior in the 1950s and early 1960s. The launch of Sputnik I established the Soviet Union as equal (if not superior) to the United States in world opinion. And the ensuing string of space spectaculars, culminating in the flight of Yuri Gagarin, created a reputation for the prowess of Soviet rocketry that was unmatched by the reality of actual missile deployments. In the classic Russian Potemkin village tradition, Khrushchev sought the appearance of strength rather than the substance of power. Nor was the symbolic potency of rockets lost on the United States, which responded with the equally symbolic Apollo lunar landing program. The achievement of an autonomous space launch capability is a no less potent symbol of national power. The first Israeli satellite launch was largely an exercise in symbolic politics, publicly displaying the potential of Israeli rocketry. Today, one measure of the status of a country as a great power is the possession of an autonomous launch capability, while a superpower is distinguished by its possession of an autonomous piloted spaceflight capability.

Maintaining a commitment of money and talent over the extended period of time required to develop an indigenous missile capability has frequently required maintenance of rocket development programs which obscure the relationship between the launch vehicles for civilian space activities, and military missiles. Tom Lehrer's observation that Wernher von Braun "aimed for the Moon and hit London" would surely find resonance with rocket development efforts in Brazil, Pakistan and India, to name but a few. In each country, civil space authorities have sponsored the development of an indigenous space launch capability, under the rationale that the nation's full exploitation of space required an autonomous space launch capability. And in each case, these long-standing civilian space exploration programs provided rocket development efforts that were converted into military missile development projects. The legitimacy of civil space exploration enabled countries to pursue missile programs for long periods, even decades.

The contention that these missiles pose a novel threat is based on the observation that they can be used to carry either chemical or nuclear warheads. But every country possessing or developing such missiles already has large numbers of aircraft with similar or greater capabilities. Some would argue that using missiles to deliver such warheads might provide an element of surprise or improve the odds of successfully reaching the target. But air defense systems have been manifestly unable to reliably detect or intercept aircraft in combat, as witnessed in the American air war over Vietnam, the 1973 Yom Kippur War, the Israeli raid on the Iraqi Osirak nuclear facility, and Soviet failings with KAL 007 and Mathius Rust. This experience suggests that aircraft may prove as surprising and reliable as missiles in delivering chemical and nuclear warheads.

During the early 1980s a few hundred missiles were fired between Iran and Iraq. Although several thousand civilians were killed, these attacks did not materially affect the course of the War. A handful of missiles were fired by Egypt at Israel in 1967, and in Afghanistan in 1989, and one missile was fired by Libya at an American base in Italy in 1987. In each case the missiles fired were SS-1 Scuds, a lineal descendent of the German V-2. During World War II Germany fired nearly 5,000 of these rockets at targets in Europe and the United

Kingdom. Although these attacks resulted in thousands of deaths and significant property damage, they were not militarily significant, and were a tremendous waste of resources from a cost-benefit standpoint. Given the marginal results obtained from the use of thousands of these rockets in the Second World War, the more recent use of much smaller numbers of these rockets should not prove surprising.

Most third world countries have followed the example of the United States and former Soviet Union by closely relating their military missile projects and civilian space programs. This approach is consistent with the overall national prestige component of both efforts. This linkage also enhances the domestic legitimacy of the military missile projects, as well as providing a rationale for rejecting international limitations on such projects.

The threat posed by these programs will increase with time. But it is also important to keep the missile proliferation issue in perspective. Longer-range missiles do not provide such unique military capabilities that nations cannot do without them. There is enough time to implement measures to control missile proliferation. The national prestige considerations that are important drivers for these programs suggest that breaking the link between space development and missile development may provide an effective means of discouraging missile proliferation.

Space Launch Systems

In the wake of the Challenger accident, the American comeback in space was marked by an expanding fleet of new and redesigned expendable launch vehicles. However, these American efforts have been marred by technical problems, schedule delays and cost overruns. The US industry has also experienced strong international competition in the commercial market. Progress toward longer-range projects, such as the National Launch System and the National Aerospace Plane, has slowed significantly during recent years. In April 1992, the Commerce Department's Commercial Space Transportation Advisory Committee recommended a program for upgrading the existing fleet of launch vehicles. Despite its modest price tag and the great benefits of this program, no action has been taken on these recommendations.

US suppliers dominated the world market for space launches in the 1970s. But the European Ariane rocket, introduced in the early 1980s, has captured over half of the international launch market. Soon European suppliers will introduce a new rocket with improved capabilities. The Japanese have similar plans, and the non-market economies of China and Russia have entered the space-launch arena.

An important aspect of American policy relates to realistic and achievable plans for the next generation of launch vehicles. Existing rockets must be improved and fair trade agreements must be struck. But these measures may prove insufficient. Eventually new and more capable launch vehicles will be needed. Unfortunately, no clear course for the future has been charted to date.

Emerging competitors pose a serious challenge to the American launch vehicle industry. The Bush Administration failed to meet this challenge. By failing to upgrade our existing rockets, by failing to negotiate fair trade agreements with our competitors, and by failing to develop new, affordable launch vehicles for the 21st century, the Bush Administration was unable to forge a coherent strategy for this industry. The Clinton Administration has yet to place its stamp on the matter.

American Launch Systems

The Titan, Atlas and Delta are all derived from ballistic missiles that were originally developed in the 1950s. Since then, they have been considerably modified to improve their ability to launch satellites. But additional improvements are possible.

Today, there are four different US programs to develop launch vehicles. The National Launch System is intended to produce a new family of conventional rockets. The National Aerospace Plane program, sometimes referred to as the Orient Express, is intended to develop a winged, air-breathing launch vehicle that can reach orbit without needing additional booster rockets. The Delta Clipper program hopes to combine the best features of both of these programs. And efforts continue on improving the space shuttle.

Combined, these projects cost a billion dollars this year, at a time when money is difficult to find. The price tag on completing development of any one of them will be at least ten billion dollars, money that will be very difficult to find. The United States may be able to finance the development of one new launch system, but it certainly will not be able to afford them all.

In the aftermath of the Challenger accident, the US government, with the strong encouragement of the Air Force, decided to promote the development of a domestic expendable launch vehicle industry. Although many reasons have been offered, United States opposition to permitting Russian and Chinese rockets to launch American-made satellites is ultimately rooted in the fear that the lower prices charged for these rockets would undercut the American rocket industry.

Unfortunately, this policy runs counter to the interests of the American satellite industry, which does a major share of its business with international customers. Since the total value of the satellite market greatly exceeds the value of the launch vehicle market, and since American manufacturers account for a greater share of the satellite market than of the launch vehicle market, it might be argued that preference should be given to promoting international sales of American satellites over promotion of sales of American launch vehicles.

The American government was confronted with such a choice when the managers of the Australian Aussat project selected Hughes to build the satellites, but also selected the Chinese Long March booster to place the spacecraft into orbit. The Reagan Administration resolved this matter, as well as the similar Asiasat question, by negotiating a special arrangement with China that established a quota on the number of boosters the Chinese could sell to launch American satellites. But this ad-hoc arrangement, which has been put on hold in the wake of the repressive measures taken by the Chinese government in early 1989, is clearly no substitute for a more generally applicable policy.

The Space Launch Market

The primary policy initiative on space launch systems focuses on concluding effective fair trade agreements with the other countries to establish "rules of the road" covering the international marketing of launch vehicles. The Bush Administration tried to negotiate "rules of the road" with our European space competitors after 1989, but failed to produce an agreement. In that year, it concluded an agreement with the People's Republic of China which set a quota on Chinese launches of American satellites, and prohibited the Chinese from selling these launches at prices below prevailing international rates. In addition, the Chinese agreed to abide by international rules on the sales of ballistic missiles to other countries.

But the Chinese violated the terms of this agreement. They offered to sell launch services at prices substantially below world prices, and sold their military rockets to countries such as Pakistan and Syria. Compounding the problem, and ignoring protests by his own Transportation Department, President Bush permitted five additional American satellites to be launched by the Chinese. This transparently political move was intended to mollify Chinese reactions to the sale of F-16 fighters to Taiwan. Then in July 1992, in a reversal of established policy, the Bush Administration gave the go-ahead for the Russian's Proton rocket to launch an American-built communications satellite.

It fell to the Clinton Administration to begin negotiations on a marketing agreement to cover such launches. In the meantime, Russian rocket makers were trumpeting the low costs of their wares, claiming they will sell their rockets at half the prevailing world price. After seventy years of Communism, Russian aerospace managers have little understanding of such basic concepts as "price" and "cost." Nor will they until a true market economy emerges in Russia. And how could they, as long as Russian aerospace companies continue to be propped up by limitless loans from the Russian government bank?

Existing Regimes

Missile Control Regime

Although the rate at which ballistic missiles may spread often is exaggerated, it is difficult to argue that missile proliferation will enhance rather than diminish international stability. But the decadal rates of diffusion provide an opportunity for action that could forestall the eventual development of reliable, increasingly threatening weapons systems. No policy short of the use of force will dissuade determined proliferators from acquiring advanced technologies. Therefore, the goal of US policy should be to narrow the playing field, forcing the kibitzers to remain on the sidelines, leaving only committed proliferators for all to see. US policy must engage suppliers and consumers; survive shifts in the global and regional balances; integrate carrots and sticks; and incorporate prudent military planning as insurance against failure.

When examined against this standard, current US nonproliferation policy suffers from a number of serious flaws. First, it is unbalanced. US nonproliferation efforts emphasize sticks rather than carrots, and military insurance rather than diplomatic initiative. Second, it is brittle. It seeks quick-fix solutions sufficient in a relatively benign global politico-military environment but unlikely to hold fast in a strong wind. Proliferators may be momentarily forced to the sidelines but may not stay there given changes in political, military or economic circumstances. In short, the US approach is ill-suited to achieving any of the four goals outlined previously.

Several schools of thought have emerged in the US missile nonproliferation community. Some contend that future efforts should focus on the MTCR. The utility of this export control approach is challenged by advocates of an "arms control" approach, who call for extending the Soviet-American Intermediate Range Nuclear Forces (INF) agreement on a global basis. A third, and growing, school of thought concludes that such efforts are bound to fail, and that deployment of the Strategic Defense Initiative is our only hope.

Most of US missile nonproliferation policy over the past five years has focused on improving the MTCR. The first priority has been to broaden membership beyond the original seven members, on the assumption that a global MTCR is the most effective means of stopping the spread. Presently 23 countries are formal members (including the seven leading economic powers: the United States, Great Britain, France, Italy, Japan, Canada, Germany), with four other nations including Russia pledging to adhere to its export control guidelines. Even China has pledged not to sell missiles that would violate the MTCR guidelines to the Middle East.

A second approach has been to sharpen the focus of the regime. The MTCR prohibits exports of technologies that could be used in missile systems with ranges greater than 300 km for a payload of 500 kg. The MTCR has been criticized for a number of reasons, one of the most important being its restrictions only apply to missiles capable of carrying nuclear weapons, not chemical and biological weapons. At the March 1991 meeting in Tokyo, in response to the Gulf War, the members extended the current objective to lighter-weight missiles capable of carrying chemical and biological weapons. This action required lowering the range and payload thresholds which were geared to missiles capable of carrying heavier nuclear weapons.

A third effort has focused on members improving their national export restrictions. The United States and other key industrialized states have taken steps to unilaterally tighten export controls. Germany, long a source of missile technology, has recently recast its export laws to strongly monitor export licensing. The United Kingdom has taken similar steps.

On the whole, an MTCR-centered policy can provide a fairly effective stop-gap in the short run. Both Brazil and India have complained about MTCR technology controls, ostensibly because they hamper each country's civilian space programs. However, the effectiveness of an MTCR approach should not be overstated. A case in point is the collapse of the Argentine Condor program which some have claimed is the result of MTCR pressure. Export controls did play an important role in slowing and momentarily preventing Argentine acquisition of key technologies to complete the missile. But, it was US diplomatic initiatives, budgetary limits, and a change in the Argentine government which actually resulted in the program being dismantled. Without a sympathetic civilian ear and sustained outside diplomatic pressure, Argentina might have continued towards building Condor.

In the long run, the MTCR is unlikely to prove sufficiently durable to effectively deal with proliferation in good and bad times. It is nonsensical to seek global adherence to the cartel, in large part because it sets no limits on a member's missile programs. Thus, countries which are the target of the MTCR's guidelines (such as India) could at the same time belong to the regime. In any case, few third world countries are likely to join the cartel. Belonging to a cartel dominated by the "North," the purpose of which is to prevent the spread of technology to less developed countries, is unlikely to prove politically attractive. Even more important, because the MTCR is essentially an exercise in denial, it offers no economic incentives for membership nor for adherence.

While improving the MTCR is necessary, it is not sufficient. There are limits to what can be done to frustrate a state that is absolutely committed to developing an indigenous missile capability. Acquisition of ready-made technology and hardware can speed or facilitate a rocket development program.

Perhaps no state since Nazi Germany has built its rocket entirely without the help of another state. But dozens of countries can obtain the elements of a long-range rocket, given enough time, money and talent.

No discussion of missile proliferation control would be complete without a recognition of the need to address the proliferation of other delivery systems, notably attack aircraft. The current US position on Mid-East arms control of discouraging missile sales by other countries while encouraging US sales of attack aircraft is unlikely to find wide acceptance in the region.

Presently there are approximately 42,000 aircraft worldwide with ranges that match MTCR restrictions on ballistic missiles. Between 6,000 and 7,000 of those aircraft are in the developing world. Furthermore, most of these have longer ranges than 300 km and can carry payloads heavier than 500 kg and refuel and fly several missions before being shot down. When one compares what aircraft did to Iraq compared what ballistic missiles did to Israel and Saudi Arabia during the Gulf War, it is evident that the "aircraft problem is orders of magnitude larger than the "missile problem."

Much of the Chinese recalcitrance on MTCR compliance must be attributable to the illogical asymmetry of American attitudes toward missile and aircraft proliferation. It must be difficult from a Chinese perspective to understand the American insistence on limiting the spread of small numbers of relatively ineffective weapons, such as ballistic missiles, while actively encouraging the spread of large numbers of far more effective weapons, such as attack aircraft. The Chinese might be forgiven for suspecting that the fact that America exports aircraft but not ballistic missiles may have some role in this curious policy.

Missile Testing

It is important to recognize that in most cases missile proliferation is an emerging problem rather than an imminent threat. Some programs, such as the Argentine-Egyptian-Iraqi Condor, were under way for many years without leading to a single test. Those newly-developed missiles (in contrast to turn-key imports) that have begun the flight-testing phase have had only a few tests. Since American, Soviet and other missile test programs have traditionally required twenty or more flight tests before the missile is considered operational, the missile programs of such countries as India, Pakistan, and Brazil may have many years to run before producing a weapon that would inspire operational confidence.

There is a great difference between a five ton missile capable of lobbing a chemical warhead a few hundred miles and a fifty ton missile that can toss a nuclear warhead over intercontinental ranges. ICBMs require industrial and engineering capabilities that go far beyond that required for shorter range rockets. Thus, it is not surprising that few countries have expressed much interest in longer-range rockets.

The United States was able to go from testing a 300 km range missile to testing a 10,000 km range missile in less than a decade, and the Soviet Union took a little over a decade for the same achievements. But France and the United Kingdom moved much more slowly, and China needed nearly two decades to complete the process. Taking into account the time needed to go from the first test of a long range missile to having an actual operational capability suggests an even longer time period.

The United States would not be caught by surprise if any country attempted to develop a ballistic missile. Developing a missile capability takes years and requires a series of flight tests. The overall historical record suggests an average of 46 tests over a 32 month period is needed to make a missile operational. Although the number of tests required has declined over time, to a current average of about 20, the total time required for testing has increased. Even the superpowers require at least five years of testing to achieve an operational capability.

It might be argued that third world countries would be prepared to forgo extensive testing. They would do so at the risk of having an extremely unreliable missile force, however. Experience with American space launch vehicles confirms that several dozen test flights are required to achieve acceptable reliability. A missile that has not been tested extensively would not be a credible or reliable weapon. It is difficult to imagine that a third world country would expend billions of dollars on a nuclear weapons program, only to see this investment wasted on an unreliable delivery system. The few hundred million dollars that would be saved by conducting an inadequate number of tests is totally disproportionate to the value of a nuclear weapons program.

These tests are intrinsically observable events, and become moreso as the range of the missiles increases. The same US satellites that tracked the Scud launches during the Gulf War would detect missile flight tests, giving unambiguous warning of missile development by any country in the world, which would provide adequate time to devise an appropriate response. Furthermore, long range missiles using less than state-of-the-art technology would be too large to be mobile. Consequently, their deployment locations would be known from satellite surveillance.

Anti-Missile and Anti-Satellite Programs

Inspired by the apparent success of Patriot interceptors against Scud missiles during Desert Storm, and capitalizing on the political disintegration of the Soviet Union, proponents of the Strategic Defense Initiative succeeded in reversing the political fortunes of the program in 1991. Whereas 1990 had witnessed a major reduction in funding for SDI, the budget approved by Congress in 1991 more than reversed the cutbacks of the prior years. Furthermore, in a major step, the Congress endorsed the eventual deployment of a large ground-based system, starting with an initial deployment at the former Safeguard ABM site in Grand Forks, North Dakota, that would far exceed the limits imposed by the 1972 ABM Treaty. The political transformation of the former Soviet Union also led to a major evolution in US attitudes toward anti-missile systems.

The Clinton Administration has embarked on a major program aimed at establishing defenses against theater and tactical ballistic missiles. During the 1992 campaign, Clinton stated that:

> We should focus the SDI program on three more concrete goals connected to hard-headed analysis of the real threats that the United States might face in the future. First, we would develop and deploy theater-based defense systems--like Patriot and its successors--to defend US Troops and allies against the existing threat of short-range missile attack. Second, we should focus strategic defense research on a limited defense of the United States against the possibility of new ICBM threats. Such threats have not yet and may never emerge--the CIA says there will be no new ICBM threat for at least a decade. But, it is prudent to be in a position to deploy a limited defense should the need arise. Third, we should support a prudent research program on more advanced follow-on anti-missile technologies. This would ensure American technological leadership in the field, as well as preserve the option to deploy more capable systems in the future, should the need arise.

In contrast to the prior debate over SDI, which was primarily driven by questions of cost and technical feasibility, the new SDI debate largely revolves around the question of the reality and significance of the threats the system is intended to counter. Whereas initially SDI was intended to replace deterrence, and in later modifications, to enhance deterrence, supporters of ballistic missile defenses now see them as a tool to cope with the potential failure of deterrence. Prior to becoming Secretary of Defense, Les Aspin suggested that deployment of SDI might be needed in the face of the emergence of "non-deterrable threats:"

> Within the past year we have seen growing signs that some future nuclear threats may not be deterrable. Saddam Hussein is a case in point...It is difficult to say what Saddam would have done if he had completed a nuclear bomb, but his actions in the Gulf War raise

serious doubts about whether he would have been deterred from using
it.

However, the case for the existence of such non-deterrable threats is unclear.
While Iraq used both chemical weapons and ballistic missiles extensively in the
Iran-Iraq War, the non-use of chemical-armed missiles in the Gulf War may
demonstrate the effectiveness of deterrence by the threat of retaliation. The fact
that Iraq did fire conventionally-armed missiles at Israel stemmed from the
unique fact that Saddam was trying to draw Israel into the war in order to split
the coalition. But even in this effort, Iraq observed a threshold that limited its
efforts to conventional weapons.

One might question whether the United States would actually have used
nuclear weapons in response to a chemical attack. Obviously, Saddam Hussein
could not have been confident that we would not. Thus, far from constituting an
example of a non-deterrable threat, the Iraqi experience demonstrates that even in
the midst of war, even the most ruthless dictator was deterred from crossing a
threshold that could have led to massive retaliation. Far from making the case
for deploying an anti-missile system, Desert Storm confirms that such a system is
not needed.

The argument in favor of anti-missile systems to address the possibility of
non-deterrable threats is eerily reminiscent of Robert McNamara's case in 1967
for deploying the Sentinel system to defend against Red China. McNamara
rehearsed all the familiar arguments against large anti-missile systems, but in the
face of Republican pressure to deploy some sort of system, concluded that the
Chinese menace warranted deployment of a small anti-missile system.

There are a few superficial similarities between China in 1967, and potential
threats today. Then, China was engaged in instigating hostility to the US around
the world. Mao Tse-tung had declared his belief that the American nuclear
deterrent was a paper tiger, and had gone a long way toward convincing the
world that he was the leading practitioner of what Richard Nixon would later
term the "madman" theory of statecraft. Under these circumstances, China
might not be deterred from launching a nuclear strike on the United States by the
threat of retaliation that seemed to deter the Soviets. Thus an anti-missile system
might be the only means of meeting the emerging Chinese nuclear threat.

Today one searches in vain for a country with the combination of
irresponsible leadership, nuclear capability, long-range missile capability and
strategic motivation that would constitute a threat that could only be answered by
deploying an anti-missile system. Some countries may currently meet one or
more of these criteria; none meet them all.

Even if the United States were to field a limited ABM system, it is unlikely
that the possession of such a system would alter the risk-aversion of the

American national leadership. There is no prospect that any anti-missile system would be deemed so thoroughly reliable that the United States would be prepared to pose an existential threat to a nuclear armed adversary. Particularly with the end of the Cold War, America faces no potential adversary which poses an existential threat to America, or even a major threat to vital American national interests. Thus, no American leader would have reason to gamble the lives of thousands of American troops or millions of American citizens by countering an adversaries use of nuclear weapons with an imperfectly tested anti-missile system.

Anti-Satellite Weapons

New justifications for the development of anti-satellite weapons now come from the supposed danger posed by third world surveillance capabilities. Advocates of the continued need for ASATs point to the potential use or purchase of satellite intelligence services by third world countries. France is currently developing its own reconnaissance satellite, slated for launch in 1994, and Israel is reportedly working on intelligence satellites as well. By the early 21st century, a number of other countries-- including Brazil, India and Japan--could also possess military reconnaissance satellites.

Until its recent cancellation, the US Army's Kinetic Energy ASAT was the Pentagon's primary weapon under research to attack hostile satellites. This ground-based interceptor would destroy satellites by homing in on and colliding with them. The technology is similar to the anti-ballistic missile hit-to-kill interceptor which was first tested successfully in the 1984 Homing Overlay Experiment (HOE), and more recently in the Exoatmospheric Reentry Vehicle Interception System (ERIS) tests. In the FY1993 Defense Authorization Conference Report, the Congress directed the US Space Command to prepare new operational requirements for the Army KE ASAT program. Congress required that the program be reconfigured to counter the space threat posed by third world nations, rather than countering Russian satellites.

The deployment of a single ABM site would provide the United States with an inherent anti-satellite weapon capability which would be sufficient to meet any likely need. However, this course of action would make the existence of any ASAT capability contingent on the decision to go forward with the deployment of a national missile defense system. Making the availability of ASATs contingent on this development would tie the ASAT program to a program that may have a weaker justification for proceeding to deployment. The perception of threats mandating anti-satellite weapons may be stronger than any threats that would require ABM capability.

The greater US threat perception associated with third world space systems relative to third world ICBMs, and the relatively lower cost of an ASAT program, may cause some to call for proceeding with development of such a system even if a ground-based ABM system were not pursued. Unfortunately, the reorientation of the ASAT program to counter third world rather than Russian threats runs into the same stability problems as the reorientation of SDI. Some may argue that ASATs are no longer destabilizing, since third world countries are not likely to acquire them and could therefore not threaten US satellites. Conversely, ASATs may be developed to counter third world satellites, but would still threaten Russian space platforms. Russian fears of satellite vulnerability would not benefit stability, as they already worry about the loss of some of their radar early warning capabilities. If the United States goes forward with ASAT deployment, then it is conceivable that the Russians would deploy their own. Thereby, other nations would be presented with a rationale to pursue and deploy their own ASAT systems. The resulting vulnerability of US space systems would clearly outweigh the benefits of US ASAT possession.

Space Arms Control

The Anti-Ballistic Missile (ABM) Treaty of 1972 is the keystone on which treaties that reduce offensive nuclear arms have rested. The end of the Cold War has greatly eased concerns over the nuclear stalemate. But for the foreseeable future, the United States and Russia will retain large nuclear arsenals targeted at each other. Even at the conclusion of the START II reductions, these forces will rival those deployed when the SALT I agreement and the ABM Treaty were signed. Under these conditions, the original logic of the relationship between limits on offensive and defensive forces remains valid. Thus, existing constraints on the anti-missile systems must remain in place if reductions in offensive forces are to be achieved.

Present and future American (and Russian) anti-missile systems threaten the continued viability of the ABM Treaty. The development and testing of some theater missile defense (TMD) components could violate the ABM Treaty well before a decision is made to deploy extensive anti-missile systems. In the near term, American and Russia are unlikely to take the provocative step of formally abrogating the ABM Treaty. Instead, each may simply undertake activities that undermine the agreement and steadily erode its restrictions until the Treaty has lost much of its significance.

The American stance toward anti-missile systems and arms control has evolved significantly in recent years. The Missile Defense Act adopted by the Congress in 1991 calls for deploying an anti-missile system that would be "cost-effective and operationally effective and ABM Treaty compliant." However, deployment of significant strategic anti-missile systems would require revision or elimination of the 1972 Anti-Ballistic Missile Treaty. Under the terms of the ABM Treaty, any operationally effective system would violate the ABM Treaty, and any Treaty compliant system would not be operationally effective.

Although the US negotiating position in the Geneva Defense and Space Talks was previously opposed to any constraints on deployment, in 1991 the Bush Administration indicated a willingness to discuss modifications to the Treaty. The position advanced by Bush would modify the Treaty to permit full deployment of the ground and space based elements of the proposed Global Protection Against Limited Strikes (GPALS) system.

Following the Bush-Yeltsin summit in June 1991, the two sides agreed that working groups would meet to explore opportunities for cooperative efforts on ballistic missile defense and early warning. During discussions of the High Level Group on missile defenses in September 1991, the United States put forward a proposed protocol to the ABM Treaty, which would: (1) permit a nationwide defense of 6 ABM sites with 150 interceptors each; (2) remove all limits on ABM deployment and testing; (3) remove all limits on testing and deployment of ABM sensors such as Brilliant Eyes; and (4) define "strategic ballistic missile" in such a way as to permit much more capable anti-tactical missile defenses such as the Theater High-Altitude Area Defense (THAAD) system. The treaty's ban on deployment of space-based ABM interceptors ("Brilliant Pebbles") would stay in place, but would expire after 10 years.

During the 1992 campaign candidate Bill Clinton endorsed continued compliance with the ABM Treaty. However, in 1993, senior officials in the Clinton Administration reversed this commitment, contending either that revisions to the Treaty are needed, or denying that systems of dubious status under the Treaty are in fact a problem.

In late 1993 the Clinton Administration proposed revisions to the ABM Treaty which would significantly expand the definition of anti-missile systems which are not subject to the testing and deployment limits of the Treaty. Over the past two decades, two tests were applied. The first test related to whether a device was tested in an ABM mode, against strategic ballistic missile targets. The second was whether a device, such as an interceptor, was capable of substituting for a strategic ABM interceptor.

The Administration proposed redefinition of the speed of a treaty-accountable interceptor from 2 km/sec to 5 km/sec, and called for the elimination of the second test, since it is explicitly acknowledged that systems such as THAAD would have a significant capability against 7 km/sec targets, even if they are not tested against such targets. But both American and Russian force planners would recognize that such a capability could be quickly demonstrated.

These revisions would effectively eliminate the present distinction between tactical and strategic anti-missile systems, and thus substantially erode the security benefits of the Treaty. Since the new interceptor systems thus freed from the Treaty's limitation would not provide a corresponding improvement in American national security, the proposed revisions are not in America's national interest. This proposed revision so fundamentally alters the ABM Treaty that its acceptance would require the advice and consent of the United States Senate. In the event that Russia acquiesced to this proposal, it should be rejected by the Senate.

Conclusions

Ultimately, the success of efforts to control the spread of missiles will depend on the success of efforts to ameliorate the domestic and international political insecurities that are the well-spring of all weapons programs. For this reason, it will be necessary to share prestige in space if states are to be persuaded to forgo self-reliant capabilities. This outcome can be accomplished by allowing access or association with the technology and activities of spacefaring nations.

National prestige considerations could be met by offering turnkey launch services, in which a complete launch vehicle would be transported to a third world country (perhaps painted in the local national colors), and launched, while remaining in the custody of the space-faring country's personnel. Currently Brazil is negotiating with the United States and Russia for such an arrangement, with Russian proposals for submarine-launched or air-launched boosters providing the greatest assurance against technology transfer.

With the adoption of this new regime, once states agreed to forgo the testing and operation of ballistic missiles, their status as weapons of prestige will be eliminated. Instead, cooperation in civil space exploration and development would substitute as the talisman of global prestige. With ballistic missiles sufficiently demystified, the proliferation problems of other sophisticated weapons may become less intractable.

Just as rocketry was a defining technology of the Cold War, it has the potential to be a defining technology of the Post-Cold War era. The long range rocket was the technological innovation that shattered the geo-political foundations of American isolation. The rocket was the defining artifact of the Cold War, both the nuclear tipped missiles of the arms race and the spacecraft of the space race. As the competition in rockets in arms race and space race defined the Cold War, cooperation in space exploration and development may become a defining activity of the coming millennium.

3

Space Power Interests: Russia

Maxim V. Tarasenko

Russia is an important element in any discussion of the long term prospects for the missile and space technology of established space powers. But Russia is a new space power with only a few years of its own history. An analysis of Russian missile and space technology and policy, therefore, must take into account the legacy of the Former Soviet Union (FSU). Current relations between the former Soviet republics also must be considered.

Because political developments in Russia and the FSU are not predictable-- even in the immediate future--it is useless to make long term projections. Hence, I have limited this study to the different trends and possible situations which could result from only one future scenario of the FSU.

In this paper, I first discuss Soviet missile and space technology and related infrastructure before and after the break-up of the Union of Soviet Socialist Republics (USSR). Then I review the status of and prospects for new missile developments and space activities. Russia would both win and lose from a regime that gives potential missile proliferators an incentive to find peaceful applications for space technologies in a new non-proliferation regime. I analyze this cost-benefit calculus on the basis of Russia's missile and space industry and its general attitude to missile proliferation. I conclude with some projections as to Russia's possible reaction to proposals for a broad non-proliferation regime and possible obstacles to its implementation.

Space Capabilities as Outgrowth of the "Rocket Shield"

The Soviet Union was the first country not only to understand the potential of long-range missiles as delivery means, but also to start actual development of an ICBM.[1]

Soviet ICBMs were developed as means of deterrence against perceived "aggressive intentions" of the United States. Since the Soviet Union could neither create strategic bombers to reach the US mainland nor had forward bases near American borders, missiles were the chosen alternative.

Although the practical military value of the first Soviet R-7 (SS-6) ICBMs was insignificant, it did become a potent tool for political pressure. The creation of the ICBM also allowed the Soviet Union to begin launches into space, which were used first as a visible demonstration of Soviet missile capabilities, and then as an additional propaganda means to show "the superiority of socialism over capitalism."

Thus missile and space technology in the Soviet Union from its very early stages served the same purpose as that of modern missile proliferators. Rockets were considered a means of deterrence and a counter to a perceived threat.

It is prudent to recall that the initial imbalance of nuclear forces rapidly evolved into a full scale confrontation between the superpowers which posed the risk of mutual assured destruction. It is useful to keep this history in mind when considering such developments in other potential confrontations.

Status of Rocket and Space Industry and Policy

The Soviet space program was one of the two biggest in the world. The Soviet Union and the United States were and are the only countries in the world to pursue the entire spectrum of space research and its scientific, economic and military applications.

The Soviet Union devoted tremendous efforts and resources to becoming a leading missile and space power. Unable to sustain total parity with the United States, the Soviet Union tried to maintain military equivalence, especially with rocket technology and state-of-the-art launchers developed both for combat and space applications.

The Soviet Union had the most missiles in the world. The Soviet Union developed about 20 types of ICBMs and IRBMs and nearly 10 types of SLBMs. Up to 1,600 ICBM and 1,000 SLBM launchers were deployed during the mid-1970s. At that time the United States had 1,200 ICBMs and less than 600 SLBMs.[2]

Having developed rocket technology of the highest standards, the Soviet Union was unable to maintain the leading edge in related fields. As a result, Soviet spacecraft were more specialized and less durable than their American counterparts. In order to solve orbital and technological problems, the Soviet Union had to launch more than double the satellites of the United States. In the 1970s and 1980s, the Soviet Union conducted 90 to 100 launches annually. This intense pace demanded production-line manufacturing of space launchers and an extensive launch support infrastructure, making the Soviet Union the world's biggest rocket power.

During nearly 50 years of long range missile development, the Soviet Union established a complex research base, a robust rocket industry, and a diverse support infrastructure including test ranges and a tracking network. Specialized space-related components of the industrial and research complex, spacecraft control and outer space monitoring networks were created, along with dual purpose or dedicated infrastructure for missile research, development and deployment.

Industry

A special department called the Ministry of General Machine-building (MOM) was in charge of the missile and space industry. This department was under the direct supervision of the Secretary of the Communist Party responsible for the military-industrial complex. No official statistics for the rocket industry were ever published in the Soviet Union. However indirect data showed that there were hundreds of enterprises under the direct supervision of MOM, working mostly for missile and space developments. The cumulative workforce was more than a million employees.

After the Soviet Union broke up, the rocket industry was divided between the republics into acutely uneven parts. The overwhelming share stayed in Russia, while some key components remained in Ukraine and the rest in Belarus, the Baltic states, and Uzbekistan. The range of estimates for the Russian share is from 75 percent of "space related properties" (measured by the value of basic funds) to 90 percent of the "enterprises."[3]

Industrial facilities in Ukraine include the largest ICBM production plant in the world and the primary production site for missile and space launchers guidance systems. Belarus retains the only industrial facility for neutralization of a highly volatile UDMH rocket fuel.

In Russia, the Department of General Machine-Building, which belongs to the new Russian Ministry of Industries, took responsibility for 214 "structural units," including 13 Production Associations, 13 Scientific and Production Associations, 20 Scientific Research Institutes, 21 Design Bureaus, 3 Scientific Technological Centers and 53 separate enterprises.[4] Total employment in 1992 was about eight hundred thousand people. The exact figure is unavailable because of the increasing problem of personnel drain. In 1992, Russian missile and space industry lost about eighty thousand employees, including 10 percent of its production workforce and 30 percent of its research staff.[5]

Support Infrastructure

The network of test ranges and tracking, telemetry and control stations was established across the Soviet Union for missile and later space launcher development, testing, and operations.

There are four known rocket test ranges:

- Kapustin Yar is in the Astrakhan' region near the Volga river. It was established in 1946 for long range missile testing. From 1962 to 1984 it was also used for launching small satellites by IRBM-derived launchers. It was later used only for surface-to-air missile testing.
- Tyuratam/Baikonur is in the Kzyl-Orda region of Kazakhstan. It was built from 1955-1957 for ICBM testing. From 1957 it was also used for space launches. This site is where all Soviet launches to a geostationary orbit took place, as well as for all manned and interplanetary missions.
- Plesetsk is in the Archangelsk region, in the northern part of Russia. It was created in 1957 for the operational deployment of R-7 (SS-6) ICBMs. From 1966 the Plesetsk site was used for space launches, mostly military. The range was later expanded to allow new ICBM testing. The SLBM test site is also situated in the Archangelsk region.
- Sary-Shagan is in Kazakhstan. It was established in the early 1960s for ABM system testing.

The tracking, telemetry, and control (TT&C) network included about 10 ground stations and a number of ship-borne stations. With the break-up of the Soviet Union this support infrastructure was also divided. The main part remained in Russia. However the key spaceport, an ICBM test site, as well as the only ABM test range are in Kazakhstan. The Russian military retains de-facto control over these facilities, which never were under local administration. However, the issue of jurisdiction over the range is a constant point of disagreement in Russian-Kazakh relations.

Ukraine is keeping a major tracking and control facility in Crimea, which greatly enhances the coverage of the entire network. Tensions between Russia and Ukraine have already resulted in removing those facilities from the Russian space TT&C network in 1992.

ICBM ground tracks neatly go through the Russian mainland to Kamchatka or the Pacific, thus remaining on Russian territory. Consequently the ICBM testing program was less influenced by the division of the infrastructure than the other space programs.

Missile Developments

The government's demand for new missile development and production began to decrease in the last years of the Soviet Union due to the decrease in the intensity of confrontation between the superpowers and the beginning of the disarmament process.

After the break-up of the Soviet Union, programs for ICBM development and modernization were cut even more drastically. Of three ICBM modernization programs already started--that of the SS-18, SS-24 and SS-25--the first two were canceled. The SLBM development program was limited to the modernization of the SS-N-20.[6]

This choice is driven by clear economic and political factors. The primary contractor for the SS-18 and the SS-24, Yangel Design Bureau and Yuzhnyi Mechanical Plant, have remained in Ukraine. Nadiradze's Moscow Institute of Thermal Technology, which designed the SS-25, and Makeev Design Bureau, which designed SLBMs exclusively, were the only dedicated enterprises in Russia which required missions. (Design bureau Salyut, which designed the SS-19, had space developments as an alternative field. Moreover, modernization of the SS-19 apparently had been rejected long before, perhaps due to design flaws revealed during operational testing). Moreover, the location of missile test facilities may have affected this decision. Test facilities for Nadiradze's and Makeev's missiles are in northern Russia, while Yangel's SS-18 and Chelomei's SS-19 have test facilities at Tyuratam in Kazakhstan.

Missile Testing

The slowdown of missile development can be observed also from the decrease in missile testing.

Historically, the intensity of testing reflected all stages of development and successive modernizations of the strategic nuclear forces. In the early 1960s the SS-7 and SS-8 first generation ICBMs were tested. In the mid-sixties, second generation (the SS-9 through SS-13 missiles) followed suit. Testing of the third generation missiles to replace the SS-9 and SS-11 (the SS-17, -18 and -19) began in 1972. The last stage in the history of Soviet ICBM testing started in 1982-1983 with the development of the mobile solid-propellant missiles SS-24 and SS-25.

A replacement for the then-obsolete SS-17, SS-18 and SS-19 was planned to start in 1997, to be followed by replacements for the SS-24 and the SS-25.[7] The modernization activities initiated would have resulted in a new round of missile testing in the early 1990s.

However, the disarmament process led to an increasing number of missile test launches since 1991 devoted to the development of conversion applications for retired ICBMs--primarily launching microgravity missions to suborbital trajectory or delivering payloads into an Earth orbit.

R&D on these applications was initiated during the Soviet era, as a test run for START-1 treaty implementation.   Later, in 1992, the new Russian administration confirmed its commitment to this approach. A government decree endorsed a "conversion" use for missiles being decommissioned under terms of START treaties. This policy provided missile manufacturers and users at least with political backing, if not with direct financial support.

Conversion applications are now being developed for practically every missile currently on duty or in storage.

- The SS-19 of Chelomei was test-fired from Baikonur on November 20 and December 20, 1991, reportedly to test a new third stage for payload in-orbit insertion.
- The Makeev's SS-N-6 and the SS-N-8 were submarine launched in 1991 and 1992 for testing a microgravity applications flight profile. The second launch, on December 7, 1992, carried a US commercial payload involving protein separation for medical purposes. Another launch reportedly occurred about June 1, 1993.
- The modified SS-25 was launched from Plesetsk on March 25, 1993 to orbit an experimental communication satellite, developed by a commercial consortium of Russian defense contractors.[8]
- The Ukrainian-built SS-18 heavy ICBMs are also considered potential space launchers. A Russian enterprise, Lavochkin NPO submitted proposals to employ the SS-18 as a spacecraft.
- Ukrainian industry proposes to develop of a series of modular space launchers from standard stages of the SS-24 ICBM. A launcher called "Space Clipper" is supposed to be fired from a container dropped from the An-124 transport plane.
- Exotic proposals include the Burlak winged space launcher to be fired from the Tu-160 bomber. The project, promoted by Design Bureau Raduga, may also rely on conversion application of Soviet air-launched cruise missile developments, like the AS-X-19. It resembles the Pegasus launcher of the US Orbital Sciences Corporation.

Makeev Design Bureau also proposed to use its newest SLBM, the SS-N-23 (Shtil'), for air-drop space launch from the Il-76 cargo plane. Recently, Sea Launch Investors, an American entrepreneur group pushing the concept of a floating space launch facility, has signed an agreement to use Makeev's missiles.[9]

That missile producers are converting missiles into space launchers is a clear and understandable trend. The former demand for missiles has disappeared forever. The stock of missiles already produced and decommissioned under

START treaties provides a vast oversupply of ready launchers, while now idle production capacity provides opportunities to make even more. A sole Makeev enterprise, for example, has a stockpile of about 200 SS-N-23 missile, not counting other types. Makeev's production capacity allows up to 20 SS-N-23 to be converted to space launchers annually.

The excess of launchers is so substantial that a desperate desire to utilize them sometimes produces strange proposals. In one conversion project, the TsNIIMash (the leading scientific research institute for the rocket industry) has studied using ballistic missiles for direct sensing of typhoons. In 1992, Mihail Maley, an advisor to the Russian president on conversion, spoke seriously about the possibility of using converted missiles for delivery of emergency aid to disaster areas.

In reality, the only practical way to use this stockpile is to reconfigure them for space launches in a national space program.

Space Program

The national space program continues after a period of acute uncertainty due to the transition from the Soviet Union to Russia.

In the 1990 and 1991 debates about the independence of the republics and the future of the Soviet Union, state and military leaders stressed that no single republic--not even Russia--would be able to sustain the space program of the whole Soviet Union. After it became clear in 1992 that the republics were unable to agree on workable cooperation in space activities, the space program of the former Soviet Union was taken over by Russia. This move allowed space program budgeting and management issues to be resolved before the end of 1992 and the decline in space activities stopped.

The Russian leadership admits that economic reality and a new political environment demand a radical change in space program priorities and a reconstruction of the space activities. However, it was decided at the outset to keep intact all aspects of the former Soviet space program. The aim was to preserve its industrial and scientific potential until new political and economic arrangements could be worked out to optimize the space program, and to make the reorientation to new tasks as painless as possible.

The commitment of the Russian state to space projects should not be measured only by launch rate, (an indicator frequently used to represent an intensity of space activity) but rather by the diversity of satellite constellations that are kept functional.

Russia now keeps operational more than 20 satellite systems, essentially all that existed before the break-up of the Soviet Union. Budgetary constraints mean that some satellite systems are sustained at a minimum level by extending the orbital lifetime of spacecraft and delaying replacements. At the end of 1992, the total constellation of operational Russian spacecraft consisted of about 140 satellites.[10] By the year 2000, Russia plans to exploit up to 30 satellite systems for scientific, economic and military purposes, a total orbital operational constellation of about 160-180 spacecraft.[11]

Although the government remains committed to the space program, the state will not order more space launchers until its budget stabilizes. Until then, the supply of dedicated space launchers will exceed demand.

Russia has no domestic, non-governmental space launch market. Independent commercial structures are too weak and the economic situation is too unstable for private investors to fund the long-term development of space systems. Launch vehicle manufacturers look to the international space launch market as the sole solution to their problems in the short- to medium-term. But the demand for converted missiles is already oversupplied by traditional manufacturers of dedicated space launchers. Until the national space program expands again, the manufacturers will aim to penetrate the international launch vehicle market.

Missile and Space Trade

Although the Soviet Union was a major weapons supplier, Soviet arms trade was never market oriented. Rather, it was always dominated by political imperatives. Moreover, the most advanced technology was usually withheld for security reasons.

As a result, the Soviet Union never traded missiles with a range of more than 300 kilometers, except in the 1950s when it provided China with the technology for production of the R-2 (SS-2) an R-5 (SS-3) missiles with a range of 600 and 1,200 km respectively. Later, Soviet missile export was restricted to the mobile tactical missiles Luna (FROG-7), R-17 (Scud B), and, more recently, Tochka (the SS-21). Soviet Scuds did allow Iraq and North Korea to start indigenous missile developments. Soviet-supplied Scuds, however, fell below or at the lower limit allowed by the Missile Technology Control Regime (MTCR) and well short of ICBMs or space launch vehicles.

The Russian government has retained this conservative approach to missile technology exports, despite its oversupply of missiles and desperate need for hard currency. Russia has declared that it shares the goals of the MTCR and will adhere to its principles. In January 1993, President Yeltsin issued a decree establishing essentially the same limitations for the export of missile-related materials, hardware and technology as are imposed by the MTCR members.[12]

After the Gulf War, the demand for missiles with anti-tactical missile defense capabilities rose sharply. Consequently, Russian missile trade is now focused on promoting advanced surface-to-air missiles, like the S-300 (SA-10) and S-300V (SA-12).

The Soviet Union never entered into the space technology trade. The FSU tried to penetrate the Western space market in the 1980s but failed. Today, Russia offers primarily space launch services and/or off-the-shelf space systems, rather than transferring technologies.

A limited transfer of space technology did occur under a cooperative scientific space program with Eastern European countries in the 1970s and '80s. The Soviet Union also assisted technically the Indian effort to develop its first satellites in the 1970s and early 1980s. But when India needed operational application satellites, it purchased them from a US company. India also acquired technology for liquid rocket engines using storable propellants (now restricted by the MTCR) not from the Soviet Union, but from France. Apparently India feared that the Soviet Union might reject a request for this technology despite the good political relations between the two countries.

In 1991 the Soviet Union proposed to build a rocket propellant production plant in Brazil as a part of an offer of a Soviet launch vehicle for orbiting the first Brazilian-made satellite. That deal would have violated the MTCR. The deal never materialized.

The Russian Attitude Toward a Possible Non-proliferation Regime

In this section, I consider how different domestic outcomes would affect Russia's response to a global non-proliferation regime. In the short term, Russia may become:

- stable internally with gradually advancing reforms;
- unstable due to a reactionary overthrow of progressive programs;
- completely unstable marked by eruption of civil war.

It is impossible to attach probabilities to each outcome. Only the first scenario provides a basis on which to make a projection about Russian policy. This scenario is assumed in the analysis which follows:

Russian Attitude Towards Non-proliferation

Russia clearly favors non-proliferation, since it prevents Russia from facing new adversaries equipped with missiles. In some senses, Russia is even more interested than the United States in non-proliferation of delivery means. An overwhelming majority of potential proliferators are close to Russian borders and the potential threat might be even bigger for Russia than it is for the United States.

A potential threat depends not only on geographical position, but on the international political posture of a country as well. If Russia abandoned its great power activities that dominate small proximate states, then the latter might treat Russia as less of a threat. However, the Russian leadership is unlikely to rely on this possibility.

The existing Missile Technology Control Regime has substantial shortcomings from the Russian standpoint, notably its discriminatory approach, which allows the MTCR to be abused to protect markets. Russia has already experienced the drawbacks of the MTCR, when in May 1992 the US State Department applied limited trade sanctions to the Russian Glavkosmos service for its agreement to transfer rocket engine technology to India. The contract had been signed by Glavkosmos (then of the Soviet Union) and the Indian Space Research Organization (ISRO) in January 1991 to supply ISRO with two hydrogen-oxygen kick motors to allow Indian indigenous remote sensing satellites to be inserted into geostationary orbit.

Russia has already suffered from inconsistencies in the MTCR. Moreover, Russia is not eligible for formal membership in the MTCR. Yet more than any other missile and space power Russia is interested in creating a new, more international and equitable non-proliferation regime.

How might Russia be affected by such a regime, one providing third world countries with access to space as an incentive for missile non-proliferation?

Such a proposal might be offered in two forms, "strong" and "weak." A strong one would provide space launch capabilities of established space powers to third world countries in exchange for their complete renunciation of indigenous rocket developments. A weak one would provide third world countries with space launch capabilities to mitigate their interest in creating launch vehicles of their own. The first form gives more confidence to rocket powers that the goals of a new regime would actually be reached, although it raises the problem of control and verification. The second, softer form is more acceptable to potential proliferators. Perhaps it would be worth trying to move incrementally from the weak to the strong regime.

Possible Gains

The proposal to provide third world countries with access to the existing space launch capabilities should receive a positive response from the Russian missile and space industry. It would enlarge the circle of potential clients and, in turn, the market for launchers. Russian manufacturers of both missile and space launchers would obtain quickly direct economic gains from such a regime. The approach would create a new market niche to compete for, whereas now Russia is trying to push American and European suppliers out of the crowded launch market.

In the long-term, the entry of Russian space launch vehicle manufacturers into an international market would facilitate a gradual integration of other sectors of the Russian space industry into a global system. Availability of Russian launch vehicles would stimulate foreign customers to contact other branches of the Russian space complex, including tiers that are currently second rate. This engagement would promote the reorganization of these branches, making them more competitive and market-oriented.

The flow of earned (rather than donated) funds into the rocket industry would stabilize the social environment in this huge and influential sector of the Russian economy. It would ease reorientation of rocket industry to new tasks and in this way assist reforming the Russian economy in general.

Besides offering economic advantages the soft proposal also promises long-term "peace dividends" by slowing down a proliferation of delivery means for weapons of mass destruction. In fact, this goal is the main reason to pursue this proposal and will be discussed later on.

Likely Doubts

A rocket power such as Russia, would ask (1) what would guarantee a country's compliance with its pledge of non-proliferation; and (2) how effective would the regime be in diverting these countries from indigenous development of launch vehicles?

Possible Losses. By providing space launch capabilities to third world countries, Russia would promote a relative decrease in its own dominance in space. This decrease itself in no way diminishes actual national space capabilities. It is only a public and political image as "the foremost space power" that would be lost. In reality, it is no loss at all, nor should it be treated as such by the current Russian leadership, which is facing a crisis of national survival.

However, the proposal would cause rightist nationalist politicians to object to a transforming Russia from "the great space power" to "cab driver for the Third World."

The related objection that once given access to space, third world countries could develop space systems that could threaten Russia (as well as other space powers) is much more serious.

Currently, only the United States and Russia have diverse systems of space surveillance, communication and control, which enhance the efficiency of military operations.

Russia's unwillingness to widen the circle of countries having similar capabilities was clearly illustrated by a reluctance of the Russian Ministry of Defense to provide access to operational military space communication systems to other former Soviet republics.

The proposal to provide third world countries with launch services would raise the issue of verification and/or restriction of satellite missions.

According to a recent statement, the US Air Force would like to have the opportunity to deny undesirable operations of space systems.[13]   The Russian space doctrine also mentions "restraining other countries from placing weapons of mass destruction in space."[14] This statement implies that Russia is able to prevent deployment of undesirable systems in space.

If the existing space superpowers declare that they intend to keep space off limits to everyone's military but their own, third world countries definitely will not give up the development of independent space launch capabilities. Western Europe's fight for an indigenous space launch capability is a case in point. Although only commercial satellite applications were at stake, Europe remained committed to developing an independent space launch capability.

To effectively restrain third world countries from national rocket developments, it will be imperative to agree on universal, non-discriminatory rules as to what kinds of activity in outer space are allowed, and what are not. The list of systems allowed for development and deployment in space must be common for all countries.   Otherwise attempts to restrain development of national launch capabilities by offering international space launch services would be fruitless, and may even fuel attempts to circumvent controls.

Such a list should include systems for military surveillance and communications and control of the sort that Russia and the United States already possess and would not give up in any case. Similar tasks are already being partially carried out by commercial communications and remote sensing satellite systems, which are already accessible to third world countries. Although the boundary between military and civil versions of these systems can be defined carefully at the outset, the distinction will be erased over time as the technology arrives.   Under these circumstances, the separation of "acceptable" and "unacceptable" activities will be largely voluntary. Attempts to restrict and control third world countries in this area of space applications would result in counterproductive conflicts. Attempts or even plans to deny some kind of space operations to these countries could also initiate a race of antisatellite weapons on their part.

The development of space-based and antisatellite weapons is the only area of space activities that definitely should be banned. Neither Russia nor the United States currently operates space weapons systems.[15] They would also disapprove of any other country obtaining space warfare capability and posing a threat to their extensive space assets. In both countries there is also a limited political base for pushing a ban on space weapons, although the odds seem long for this undertaking to be successful.

If a universal standard of permitted space activity can be defined and accepted, then the problem of enforcing compliance is the same for every country. Enforcement per se has no direct relation to a particular regime of providing launch services for third world countries. Furthermore, such a regime could be reliably verified by non-intrusive means, that is, by remote monitoring to assure that weapons are not tested in space or against space-based objects.

This approach would avoid the problem of direct inspection or control of third world countries' space objects to be launched, giving third world countries confidence in their freedom of action within commonly accepted limits. Hence, their motive to develop indigenous launchers for space transportation needs would be reduced and the effectiveness of the non-proliferation regime would be enhanced.

It would be much easier for third world countries to accept a strong non-proliferation regime, that is "access to space in exchange for a guaranteed renunciation of rocket developments," if the superpowers themselves gave up ICBMs as a delivery means.

The development of alternative nuclear delivery means, like long range cruise missiles and stealth bombers, suggests strongly that in the long run ICBMs will become much less important. Moreover, advances in computer and guidance technology shifts the focus of weapons developments from refining ICBMs to designing highly precise conventional weapons.

For Russia, however, complete rejection of ICBMs is inconceivable in the foreseeable future. ICBMs constitute the main component of the Russian Strategic Deterrence Force. ICBMs will remain a backbone of the Russian strategic arsenal for as long as Russia relies on nuclear deterrence and cannot afford to restructure its deterrent forces.

Influential conservative sectors of the Russian polity would view moves to "de-missile" Russia as an attack on its international prestige. Many politicians and the military believe that missiles are a form of great power currency and are virtually the only asset inherited from the FSU. To support the softer proposal in the short run, Russia could accept some measures for missile testing limitations. This step would build confidence on the part of third world countries that by the superpowers are willing to make further cuts in their missile capabilities. This step would not be difficult for Russia, since ICBM development and production are declining already because of economic problems.

Consequently, the current difficult situation in Russia and the former Soviet republics offers a unique opportunity to implement a new regime that would offer access to space to non-ICBM-capable states in exchange for non-proliferation commitments. Russian-launched electronic intelligence satellites, for example, are reportedly manufactured by Ukraine, which now naturally falls into a "third countries" category. Ukraine, as a non-nuclear state, abandoned production of ICBMs at the former Union facilities.

As an initial step in the practical implementation of a new regime, it would be advisable to establish an international agency to provide space launch services to third world countries. The agency should not be administered exclusively by Russia and the United States, though rocket powers should retain full operational control of launch vehicles involved. The eligibility of third world countries for these launch services might be tied to their compliance with missile non-proliferation, judged by an international administration (acting like the International Atomic Energy Agency does in developing nuclear fuel cycle capabilities under the Non-Proliferation Treaty).

In this context, decommissioned ICBMs should be used as the primary means to launch third world countries' payloads into space. This approach would be politically beneficial and would relax tensions between Russian and American launch vehicle manufacturers on the issue of Russian dumping in the market. Under the proposed scenario, converted launchers (which are cheap due to substantial governmental pre-investment which is not fully reflected in their current market launch prices) would be offered to clients who otherwise would be priced out of the market for space launch services.

Russia and Kazakhstan would probably agree to establish a center for space launch servicing of third world countries on the basis of the Baikonur launch site. This role would not violate the spirit of non-proliferation, since all facilities for shooting the missiles in question are already available at the site.

The approach of using decommissioned missiles could be prolonged easily, if after complete fulfillment of the START-1 and START-2 treaties, the United States and Russia agree to reduce further their remaining ICBM forces.

Conclusions

The idea of establishing a new international missile non-proliferation regime corresponds to Russia's general policy for the non-proliferation of delivery systems for weapons of mass destruction. Russia might provide strong support for a practical version of this concept because it could enhance international security. The proposal to provide potential proliferators with access to space as an incentive for non-proliferation promises direct economic benefits to Russia both in the near term via use of excessive capabilities and in the long run due to the integration of Russia's rocket and space industry into the world system.

Russia has little, if anything, to lose as a result of a new regime that rests more on incentives and less on the controls of MTCR. However, there are important questions to resolve before policymakers in Russia will be persuaded that a new regime is feasible and beneficial.

Further study of the reformed missile non-proliferation regime should be conducted jointly with the potential proliferating states to work out precise formulae and procedures, which would allow the regime to be implemented in practice. In this regard, the most important issues are: (1) the principle of access to space launch services; (2) the obligations of the parties and guarantees of compliance; and (3) a universal list of space activities that are allowed and prohibited for all countries. It is likely that this list would allow "force enhancement" systems for terrestrial warfare but would ban space-based and anti-satellite weapons for space and terrestrial warfare alike.

Notes

1. The Soviet Union initiated ICBM development in the early 1950s. The United States started preliminary research as early as 1947 but gave no priority to missile development until 1955, after significant Soviet progress was reported.

2. Thomas B. Cochran, et al. "Soviet Nuclear Weapons," *Nuclear Weapons Databook*, vol. 4 (Harper & Row, 1989), p. 100; "US Nuclear Forces and Capabilities" *Nuclear Weapons Databook,* vol. 1 (Ballinger, 1984), p. 101, 102.

3. Nezavisimaya Gazeta, 25 September 1991, p.6; *Komsomol'skaya Pravda*, 8 October 1991, p. 2.

4. Supreme Soviet of Russian Federation, 1992; in late 1992 the Department of Industries was abandoned and partially replaced by the Committee on Defense Branches of Industry.

5. Ibid.

6. Izvestia, April 10, 1993, p. 15.

7. Ibid.

8. This launch was declared to be the first practical Russian experience of using a former combat missile for a space launch. However, decommissioned combat SS-s were reportedly used for space launches long before. The statement might be a media trick, implying "the first missile, built in Russia." At any rate, this is definitely the first case of a former military missile launching into orbit a *non-military* satellite.

9. "US Entrepreneurs Seek Russian SLBMs," Aviation Week & Space Technology, April 19, 1993, p. 22.

10. Yu. Koptev, interview to Moscow Radio, January 4, 1993. See also *Space News*, January 11-17, 1993, p. 11.

11. "Space Program of the Russian Federation up to Year 2000," (draft) Russian Space Agency, 1992.

12. Yeltsin Issues Order on Missile Technology Exports, FBIS-SOV-93-012, January 21, 1993, p.45.

13. "USAF Chief Calls for Space Defense Upgrades," *Space News*, April 19-25, 1993, p. 1.

Maxim V. Tarasenko

14. Supreme Soviet of Russian Federation, 1992.

15. The Soviet Union deployed a co-orbital ASAT system in 1976. There was no report on decommissioning the system, although it has not been tested since 1982.

4

China's Space Interests
and Missile Technology Controls

Yanping Chen

In 1987, seven countries initiated the Missile Technology Control Regime (MTCR). The same year, China launched its commercial venture of high technology products driven by the country's economic reform. Space technology products constitute a relatively large proportion of these high technology products. These space products include the Long March launchers and missiles. The space technology business is sensitive because it involves issues of military technology transfer and also challenges the market dominance of the big players.

It is not surprising that the Chinese business practices in space technology have bothered major Western space powers. The governments in the West began to intervene in Chinese business practices with bilateral talks and by imposing sanctions. In dealing with Chinese offerings of launch services, Western governments focused on technology transfer, international treaties,and market share (in other word, quotas). In the missile trade, the intervention mainly functioned to depress trade, using non-proliferation as an excuse for sanctions.[1]

This paper offers an integrated treatment of Chinese civilian and military space products. It presents a basic understanding of why the Chinese have commercialized their space products in the international market and it suggests how the Chinese might respond to an initiative for a non-proliferation regime in missile technology.

In this paper I first describe briefly China's space capability in ballistic missiles and civilian launchers, including testing and trade. Next, I analyze Chinese motivations for developing rocket technology during the Cold War era as well as commercializing space products during the early 1980s. I include basic historical background that will put commercialization in the larger perspective of Chinese economic reform. Finally, I summarize the Chinese government's policies and attitudes towards international space treaties and the MTCR. I conclude by projecting how China may react to any new initiatives in missile technology control.

China's Rocket Technology, Test and Trade Activities

China Space Launch Capability

China began to develop its contemporary rocket technology in 1957. China has possessed 12 models of ballistic missiles for military use and 7 models for civilian launchers.

Chinese ballistic missiles are mainly of the Dongfeng series, including DF-1, DF-2, DF-3, DF-4, DF-5, DF-21(JL-1), DF-31(JL-2), DF-41, DF-25, DF-15(M-9), DF-11(M-11) and 8610. Table 4.1 shows that thirteen years elapsed between the time China began to work on space technology and the first successful test of its Inter-Continental Ballistic Missile (ICBM). China spent eleven more years to develop a submarine launched ballistic missiles (SLBM). China achieved such missile capacity with little foreign assistance.

The Long March (LM) series of civilian rockets has 7 models: LM-1D, LM-2C, LM-2E, LM-3, LM-3A, LM-4 and LM-2E/HO (see Table 4.2). Chinese launchers have developed the full capability to launch an object into three orbits: low earth, polar and geostationary. The Long March rockets are the only vehicles for domestically made satellites. China is also marketing these rockets in the international market. Since the China Great Wall Industry Corporation began to market the Long March in 1985, three foreign satellites have been launched aboard Chinese rockets.

Although China's space program is only moderate in scale, its rocket technology is fairly impressive. China is the fifth entity (after the former Soviet Union, the United States, Japan and the European space consortium) to launch a satellite; the fourth after the former Soviet Union, United States, and European Space Agency to launch multiple satellites using a single rocket; the third country after the United States and France to launch a rocket with high-energy cryogenic fuel; and one of five countries to launch geostationary satellites.[2] Given such space abilities, China is an important player in the international space club.

Table 4.1 Chinese Ballistic Missiles

	Range (km)	Payload (kg)	Technical	Availability
DF-1 (1059;SS-2)	590	950	Single stage; non-storable liquid fuel	First successful test on Nov. 5, 1960. Deployed 1961-1966
DF-2 (DF-2A;CSS-1)	1,050-1,250	1,500	Single stage; non-storable liquid fuel	First successful test on June 29, 1964. Deployed 1966-1979
DF-3 (DF-3A;CSS-2)	2,650-2,800	2,150	Single stage; storable liquid fuel	First successful test on Dec. 26,1966. Deployed 1971-
DF-4 (CSS-3)	4,750	2,200	Two-stage; storable liquid fuel	First successful test on Jan. 30,1970. Deployed 1980-. Converted into LM-1
DF-5 (DF-5A;CSS-4)	12,000-13,000	3,200	Two-stage; Storable liquid fuel:N2O4/UDMH	First successful test on Sept. 10, 1971. Deployed 1981-. Converted into LM-2 and other series
JL-1/DF-21 (DF-21A;CSS-N-3)	1,700-1,800	600	Two-stage; Solid fuel	JL-1 is submarine-launched, first successful test on Oct. 12, 1982; DF-21 is land-mobile, first successful test on May 20, 1985. Both operational
JL-2/DF-31	8,000	700	Three-stage; Solid fuel	JL-2 is SLBM; DF-31 is land-mobile. Both are expected to be operational in mid-1990s.
DF-41	12,000	800	Three-stage; solid fuel	expected to be operational in late 1990s.
DF-25	1,700	2,000	Two-stage; solid fuel	land-mobile modification from DF-31. Expected to be operational in mid-1990s.
DF-15/M-9	600	500	Single stage; solid fuel	Exhibited in Nov., 1986. First successful test in June 1988. Code DF-15 for domestic use; Code M-9 for export.
DF-11/M-11	300	500	Two stage; solid fuel.	a photograph was displayed at an exhibition in 1988.
8610	300	500	Two-stage; solid fuel	Modification from HQ-2 surface-to-air missile.

Source: J. W. Lewis and Hua Di, "China's Ballistic Missile Programs, Technology, Strategies, Goals," International Security, Fall 1992, Vol. 17, No. 2, pp 5-40.

Table 4.2 Chinese Civilian Launchers, Long March Family

	Mission	Payload Mass	Technical	Avail.	Note
LM-1D	LEO	750 kg	Three-stage launcher, Solid fuel for 3rd stage; converted from DF-4, which is a two stage missile	1970	
LM-2C	LEO	2,800 kg	Two-stage launcher, liquid fuel, converted from DF-5	1975	LEO capacity can be 3,000kg. With 3rd stage, the capacity can be reached GTO 1,000kg
LM-2E	LED	9,000 kg	Two stage launcher with strap-on engines	1991	With 3rd stage, the capability can be reached to GTO 3150kg
LM-3	GTO	1,450 kg	Three stage with liquid fuel	1984	
LM-3A	GTO	2,300kg	Three stage with liquid fuel	1993	
LM-4	SSO,GTO	SSO-2,500kg, GTO-1,000kg	Three stage with liquid fuel	1988	
LM-2E/HO	GTO	4,800 kg	Three stage with liquid fuel with strap-on engines	1994	

Source: China Academy of Launch Vehicle Technology, China Academy of Launch Vehicle Technology, Beijing, 1991.

The Connection Between Missiles and Launchers

Technically, the only major differences between a ballistic missile and a civilian launch vehicle are the trajectory and the payload. Once a country possesses Intermediate Range Ballistic Missiles (IRBM), it possesses the capability to place a satellite in a low orbit. The differences are associated with intentions rather than technical capability.[3] Chinese civilian launchers evolved from missiles. The DF-4 (range 4750 km and payload 2200 kg), for example, was successfully tested on January 30, 1970. Three months later on April 24, a converted from DF-4 known as LM-1 launched the first Chinese made satellite Dongfanghong-1 (East Is Red-1) into space. Later, the DF-5 was converted into LM-2E and other civilian launchers. Since then, the nation has enjoyed a full range of missiles for both military and civilian purposes. The rockets can lift satellites into three different types of orbits.

The close connection between ballistic missiles and launchers is reflected in not only a quick technical conversion but also in the organizational structure. The Ministry of Aerospace Industry is the only organization that conducts the designing, manufacturing and testing of both ballistic missiles and civilian launchers. Therefore, both military and civilian space missions are conducted through one administrative agency. Personnel involved in design have hands-on experience both in missiles and launchers. The manufacturer assumes contracts to build both missiles and launchers. This arrangement contrasts with the organizational structure in the United States in which SLVs and military missiles are separated institutionally.[4]

Missile and Launcher Flight Tests

China built its first missile flight test center in Jiuquan County, Lanzhou Province in 1958. Although originally constructed only as a test center, the facility became an operational complex that launches recoverable satellites for both military and civilian uses. Most Chinese missiles of different range have been tested at this center. Because China has a vast land area, most flight tests can be operated within the nation's border. The ICBM DF-5, tested in September 1971, was flown from this center and landed within the country's borders. In 1980, the DF-5 was successfully launched from this center to the Pacific Ocean. Underwater tests today are conducted in the Chinese territorial water.[5] Faced with a missile test ban, the Chinese would argue that testing within their border is a sovereign right of all states.

Trade of Missiles and Launchers

Before the end of the 1970s, international restrictions on missile technology transfer to selected countries (such as North Korea) were mainly part of the Cold War era and ideology. Since 1983, the People's Liberation Army (PLA) experienced a traumatic reform that involved budget cuts and personnel layoffs. The impact on the space industry was reflected in a systematic conversion from defense-oriented manufacturing to civilian work. Marketing the Long March rocket and missiles became a means to supplement the inadequate budget of both the defense industry and the Ministry of Aerospace Industry. Finding a way to support inadequate budgets impelled a series of institutional innovations.

Poly Company, for example, was established as an outlet of the Department of General Staff to begin the missile and arms trade. New Era Company is a similar enterprise set up by the Commission of Science, Technology and Industry for National Defense. The China Great Wall Industry Corporation is the sole representative of the Ministry of Aerospace Industry to provide international civilian space services.

In the 1980s, during which commercial activities gathered momentum, the supply of Chinese space products was driven mainly by market forces. In spite of the aggressive marketing by Chinese companies in space and arms markets, China is still only a peripheral technology supplier. As a result, Chinese companies have only a marginal share in both launcher and missile markets. In the space launcher market, for example, China has gained only a 4 percent share, compared with the French with 63 percent and the United States with 34 percent.[6] In the arms sale market, China gained only a 3 percent market share from 1985-1989, ranking behind the former Soviet Union, the United States, the United Kingdom, and France.[7]

Before China decided in 1992 to apply the MTCR's guidelines and parameters to its own exports, the nation had its own policy for arms sales. Government policy permitted the sale of "the arms used for defensive purpose to friendly countries."[8] This policy is still valid today, but only as a diplomatic statement. The policy is very flexible in principle. First, there are no technical data to define what kind of weapon and power are "defensive." Second, no clear definition exists on which countries are "friendly." Most of the sales decisions are made on an ad hoc basis rather than on a planned and strategic one. These broad interpretations permit the economic aspect of the sales to outweigh either consideration when China's arms export corporations are closing deals. Without government oversight or regulation in arms sales, the companies covertly sell missiles to whomever can pay for them. Consequently, Chinese business practices have clashed with American non-proliferation standards, which are rooted in American global interests.

Interestingly, lack of collaboration between China and other space powers in rocketry technology, even in basic technical terminology, can result in misconceptions in how business should be conducted. At the time China began to market its missiles, it still employed its early classification in missile ranges. In their technical terminology, only missiles with ranges above 1,000 km are considered strategic ones. Table 4.3 shows that M-9 and M-11 are considered, in the Chinese definition, as tactical missiles. Recently, I asked a Chinese missile engineer if he knew the distinction between the Chinese classification and the Western one. He responded that he did not. It is clear that the missile classification and terminology must be clarified if China is to participate in multilateral control agreements.

In the post-Cold War era, the Chinese-Western confrontation has been mainly an issue of how to conduct arms sales without promoting regional instability, rather than the continuation of an ideological struggle. Ideological criticism of Chinese business practices are not valid in the post-Cold War era. The Chinese government has been willing to make compromises in order to reduce trade tensions with Western countries. In 1992, China joined the Nuclear Non-Proliferation Treaty, becoming one of ten additional states to sign

Table 4.3 Chinese Classification in Missile Ranges

Chinese Classification	Range (km)	Western Classification
Short-range (Jingcheng)	< than 1,000	Short Range Ballistic Missile (SRBM)
Medium-range (Zhongcheng)	1,000-3,000	Medium Range Ballistic Missile (MRBM); Intermediate Range Ballistic Missile (IRBM)
Intermediate-range (Zhongyuancheng)	3,000-4,800	IRBM
Long-range (Yuancheng)	3,000-8,000	IRBM; Inter-Continental Ballistic Missile (ICBM)
Intercontinental (Zhouji)	over 8,000	ICBM

Source: Chinese classification see Chinese Encyclopedia, Zhongguo Dabaike Quanshu: Junshi Juan (Chinese Encyclopedia: Military Affairs), Chinese Encyclopedia Press, Beijing, Shanghai, pp. 504, 1219 and 1234. Western classification denotes that missile range from 200-1,000 km as SRBM, 1,000-1,500 km as MRBM; 1,500-5,000 km as IRBM and over 5,000 km as ICBM.

the treaty. In the same year, China pledged to apply the MTCR Guidelines to the transfer of missile technology, especially in regard to M-9 and M-11 missiles. These missiles had been attractive to countries such as Pakistan, Syria and Iraq. This promise was officially confirmed by a written commitment from China's Foreign Minister Qian Qisen on February 1, 1992.[9]

China's long term attitudes toward an international missile technology control regime will depend in part on the nation's motivations for developing and commercializing rocket technology. In the next section, I will discuss Chinese motivations in two distinct history periods. One is the period from the mid-1950s to the later 1970s; the other is from 1979 to the present day.

Motivations Behind Rocket Technology Development and Trade

In the Cold War Era

China began to develop its launch capability in 1957. The impact of the earlier nuclear threat to China from America during the Korean War certainly served as a driving force for the nation's strategic weapons program. The later withdrawal of Soviet assistance steered China toward rapid development of a self-reliant space industry. The development of a sophisticated space industry was seen to serve defense needs and to provide the nation with international prestige.

So far, no evidence has been released to show that China intended to develop missile technology for offensive military purposes. China had a broad military defense strategy aimed at American military bases in Asia. Later, it had a military defense strategy directed toward the former Soviet Union's offensive forces. China possessed no strategic arrangement for any offensive actions.[10] China's relatively weak conventional weapons force made it unlikely the nation would have taken any offensive action far beyond its borders.

Evidence that China developed missile technology for defensive purposes also exists in early leaders' doctrines. In 1956, Mao Zedong said that "We need to have nuclear bombs. In today's world, we cannot avoid the others' bullying without this." In 1958, he said, "I think that we can make nuclear bombs and inter-continental ballistic missiles in about ten years." Later, he pointed out that the principle of the development of China's strategic weapons is to have "a little, a few and the better ones."[11] These words clearly show that China's decision to have ballistic missiles was mainly for defense.

Beyond this basic consideration, a strong space capacity was emphasized in the early stages of the development of China's missile technology to obtain: 1) an instrument for a less compromised foreign policy; 2) a symbol of international prestige and national pride; 3) a leading force to stimulate the development of other industries; and 4) a stimulant to scientific and technological research.[12]

Many observers of China's space program marvel at how such a low-income country developed such advanced capabilities with its independent efforts. The reason for China's success is tied to certain attributes of its program, namely:

- High-level leaders in the central government have consistently offered strong, high-priority support for the space program, especially for the early missile program.
- The policy of self-reliance forced China to free itself from dependence on overseas expertise, though China welcomes outside technical assistance as appropriate.

- Development of a space program has been accompanied by the parallel building of a significant space industry to support the program.
- In its early stages, the program's build-up was tied closely to the missile program, so that advances in missile technology could translate immediately to space launching applications.
- The space program was justified in part because of the contribution it made to stimulate China's broad scientific and technological and industrial capabilities.

Economic Reform

China began to focus on economic development in 1979, well before the Cold War ended. Underlying the push for economic reform was the belief that another big world war was unlikely. Despite the need to maintain sufficient defense capacity, China believed it could not afford to lose time if it was to catch up economically with the West. China began to demobilize large numbers of military personnel and to cut military spending dramatically. The space sector was hard hit because 80 percent of its revenue came from defense production. The sector switched quickly from manufacturing defense products to civilian goods and sold space technology on the international market to earn hard currency. The key historical events that led to the commercialization of space products are as follows:

- In March 1978, Deng Xiaoping called for the defense sector to serve the economic growth of China.[13]
- In August 1978, Deng Xiaoping informed the Ministry of Astronautics Industry that the space industry must contribute to economic development. It could do this by focusing on developing satellites that would produce economic benefits for Chinese society. He stated that China has no intention to compete with the United States and then the Soviet Union to go to the moon (which meant that China will not compete with super powers on space expenditure).
- In 1981, China launched three satellites aboard one rocket into low orbit. In 1982, China successfully conducted an underwater launch. In 1984, China successfully launched a geostationary satellite. Since then, China claimed that its space sector possessed the technology capability to fully serve domestic and international demands.
- In early 1985, the Ministry of Astronautics Industry made a decision to offer Long March rockets and satellite services to the international market. The Great Wall Industry Corporation began to commercialize civilian

space products. Almost at the same time, the military enterprises in China began to market weapons, including missiles.

- In 1989, the Ministry of Aerospace Industry, a merger of the Ministry of Astronautics and the Ministry of Aeronautics, announced that a successful defense conversion brought 80 percent revenue from civilian production; and military production accounted for only 20 percent.[14]

From the above events, it is clear that China's motivation behind marketing space products is mainly driven by hard economics. Meanwhile, trading space products has also brought international prestige (visibility) to China because of sophisticated space technology.

China's Attitudes Toward International Obligations in Space and Its Viewpoint About MTCR

China and International Obligation in Space

Before the late 1970s, China was largely isolated from the international community. There are two reasons for this inactivity. First, China was rejected from the United Nations before 1972. Second, the Cultural Revolution from 1966-1976 isolated the whole society from the international community. The economic reform begun in the late 1970s also increased China's participation in international affairs, both in relation to the exercise of rights and the incurring of obligation. In the space arena, China signed three space treaties in November 1988: the Agreement on Rescuing Astronauts and Sending Them and the Object Launched into Outer Space Back to Earth; the Treaty on International Responsibility for Damage Caused by an Outer Space Object; and the Treaty on Registration of Objects to Be Launched into Outer Space.[15]

From an historical perspective, the timing of these treaties is noteworthy. China did not sign the agreements until not doing so became an obstacle to the Long March launching of American made satellites. After the Sino-US talks on October 21, 1988, the Chinese People's Congress approved a request from Prime Minister Li Peng to sign the three treaties on November 3. Thus, only 10 days passed between the Sino-US talks and China's formal approval of the three outer space treaties. By moving quickly, the government demonstrated its commitment to providing a favorable environment for space activity.

The government has no explicit policy to deal with space trading issues. Nonetheless, the Chinese government responds quickly to foreign government requests in order to support Chinese business interests. Clearly, the Chinese government still plays a significant role in international business conduct, especially in dealing with international obligations.

China and the Missile Technology Control Regime

China pledged to adhere to MTCR guidelines in return for the lifting of sanctions imposed by the United States in May 1991. These sanctions included restrictions on $330 million worth of computer technology and the launching of US satellites on Chinese rockets.[16] It is unclear how strong the Chinese government's commitment is to sign the MTCR. In addition to China's unwillingness to give up the profits of selling space technology, the ambiguous nature of MTCR itself makes it likely that China's space trade will be subject to dispute. Chinese publications have anticipated this outcome with a series of arguments.

First, the MTCR is said to be the product of the prevailing views of the United States and its Western allies. The regime is based on the United States' new global strategy to protect its global interests at the expense of third world countries that want to develop their own civilian space technology. This view will be especially significant in the domestic debate over the transfer of dual-use space technology.[17]

Second, Chinese analysts believe that the agreement may not attract most countries that have or potentially have ballistic missiles to participate because:

1. Nearly one hundred countries are already engaged in space technology research and development, of which about twenty have their own space agencies. By 1991, eighteen countries possessed civilian launch programs.[18] The interests these countries in having launch capability vary from one to another.
2. Outer space belongs to all of humanity. Every country has the right to develop space technology to make peaceful use of outer space. The MTCR is seen as serving the interests of a small number of nations using missile non-proliferation as a way to monopolize space industry and technology.[19]

Third, Chinese analysts doubt its enforceability. In the words of Ye Yuan, the MTCR "is not a document of international law with any real force. It is neither a treaty nor an administrative agreement. It provides no international organ for enforcement, nor has it any clause concerning verification or observance of its requirements."[20]

Fourth, the Chinese regard American efforts to control non-US arms sales (including missiles) to conflict regions in the world as hypocritical when--at the same time--the United States negotiates multi-billion dollar sales to the very same regions. This double standard is viewed as untenable.[21]

Conclusions

Its achievement in space technology has been a major symbol of international prestige for China. China is especially proud because this achievement was mainly derived from its independent efforts. China certainly expects to accrue prestige and profits from its space program.

China has been a very active participant in international space activities for peaceful uses. It has various space science and technology cooperation agreements with about seventy countries. In the last two years, China has been active in initiating an Asian Space Agency, which will include all Asian countries to pool resources for a strong regional civilian space program. It is very likely that China is willing to accept some approaches to promote international cooperation in various space programs in exchange for a reduction of the military use of space technology.

Based on the above information and analysis of Chinese motivations in developing and promoting rocket technology, it seems unlikely that China will commit itself to a more restrictive missile technology regime, especially one that will hurt its space industry and market shares. It is not clear that China is holding out on missile control as a bargaining chip in trade negotiations. However, it seems that an increased commitment to missile non-proliferation would entail compensating China for the resultant economic losses. This outcome might be achieved through certain trade adjustments in bilateral and multi-lateral trade relations. For example, compensation could be effected by allowing free exports of certain high technology products to meet Chinese needs in economic development.

Notes

1. For a history of Chinese business in Long March rockets see Yanping Chen, "China's Space Commercialization Effort, Organization, Policy and Strategies," Space Policy, Butterworth-Heinemann, London, February 1993, pp. 45-53. To see details on conflicts in Chinese missile trade, refer to Timothy V. McCarthy, *A Chronology of PRC Missile Trade and Developments,*

International Missile Proliferation Project, 425 Van Buren Street, Monterey, CA 93940, February 12, 1992.

2. Li Xiao, "Space Industry Joins World League," China Report, Joint Publication Research Service, JPRS-CST-90-027, October 29, 1990, pp. 5-7.

3. M. Eisenstein, "Third World Missiles and Nuclear Proliferation," The Washington Quarterly, vol. 5, no. 2 (Summer 1982), pp. 112-115.

4. Yanping Chen, "China's Space Commercialization Effort," p. 46.

5. Zhang Jun, et. al., Dangdai Zhongguo Hangtian (Contemporary China's Space Effort), Chinese Social Science Publishing House, Beijing, 1984, pp. 123, 150, 363.

6. Ibid., p. 51.

7. United States Arms Control and Disarmament Agency, World Military Expenditures and Arms Transfers, 1990, released November 1991, p. 16.

8. Telephone interview with an officer in the Office of Military Attaches in the Chinese Embassy in Washington D.C. on May 3, 1993.

9. J. A. Mores, "US Lifts Missile Sanctions on China," Foreign Policy Backgrounder, Press and Cultural Section, Embassy of the United States of America, FR-1736, March 27, 1992, pp. 1-2.

10. John W. Lewis and Hua Di, "China's Ballistic Missile Programs, Technology, Strategies, Goals," International Security, vol. 17, no. 2, Fall 1992, p. 5.

11. Yang Heng, "China's Strategic Weapon Development,"Nie Li and Huai Guomong, eds. Retrospection and Prospection, Forty Years of Defense Technology and Industry of New China, National Defense Industry Press, Beijing, August 1989, p. 158.

12. For details of Chinese motivations in developing missile technology in different time periods, see Yanping Chen, "China's Space Policy, A Historical Review," Space Policy, Butterworth-Heinemann, London, May 1991, pp. 116-128.

13. Chinese Economic Reform policy was carried out during the Third Plenary Session of the 11th Central Committee of the Chinese Communist Party, December 18-22, 1978, Beijing. This session formally announced that China's highest priority was economic development. The call for the defense sector to serve China's economic goal from Deng was during the National Science and Technology Conference, March 18-31, 1978.

14. "Status, Prospects of Space Industry Reviewed," China Report, JPRS-CST-90-017, Joint Publications Research Services, June 25, 1990, pp. 8-10.

15. Foreign Broadcasting Information Service, "Li Peng Submits Space Treaties Motion to NRC," China Report, FBIS-CHI-88-215, Nov. 7, 1988, p. 29 and "NPC Approved Li Peng's Motion on Three Treaties," China Report, FBIS-CHI-88-217, Nov. 9, 1988, p. 36.

16. Sanctions issued by the Bush Administration in June 1991 imposed a ban on twenty licenses worth $30 million of computers intended for China Precision Machinery Import-Export Corporation (CPMIEC) and another $300 million made by China Great Wall Corporation to launch three satellites made by Hughes. For details see United States Arms Control and Disarmament Agency, Annual Report to the Congress 1992, United States Arms Control and Disarmament Agency, Washington, D.C. 20451, 1992, p. 141 and T. V. McCarthy, *A Chronology of PRC Missile Trade and Developments,* footnote #1.

17. Ruan Zonghai, "Post Cold War Non-Proliferation Diplomacy," China Report, Joint Publication and Research Service, JPRS-CAR-92-066, August 31, 1991, pp. 1-2.

18. Donald J. Blersch, "Civil Programs and R&D efforts, Summary Matrix: World Space Programs and R&D Efforts," International Space Capabilities Reference, STDN91-14, Space Technology Division Note, Suite 800, 1215 Jefferson Davis Highway, Arlington, VA 22202, August 1991.

19. Ye Yuan, "Ballistic Missile Proliferation--How Do We Control It?" International Studies (Guoji Wenti Yanjiu) Beijing, no. 3, July 13, 1990, p. 34.

20. Ibid., p. 24.

21. McCarthy, *A Chronology,* p. 2.

5

Space Policy and Missile Control in Europe

Jürgen Scheffran

The Transformation of European Space and Launch Policy

Europe's space program is undergoing a rapid and radical transformation. Europe has discovered that ambitious and costly attempts for autonomy in space are difficult to realize, given the financial restrictions and problems of political legitimization. With the end of the Cold War, military incentives, which were never prominent in European space policy, are losing ground. Other space powers confront the same dynamic. But due to its multinational character, it is more difficult for Europe to adapt to this new situation.

Unlike other space powers, Europe is not a single nation but is composed of sovereign states with independent space policies (the Commonwealth of Independent States, CIS, is moving toward a similar structure). One must distinguish between truly European aims and the mere harmonization of national policies and goals.

The European Space Agency (ESA) was formed in 1976 by merging the European Scientific Research Organization (ESRO) and the European Launcher Development Organization (ELDO), which had been created in 1964. Today, ESA has 14 member nations.[1] Between 1989 and 1990 the ESA budget fell from $2.4 billion to $2.2 billion. The biggest contributors to ESA are France, Germany and Italy, which together contribute about 70 percent of the ESA budget. Compared to the United States, the ESA budget amounts to about 20 percent of the NASA budget. However, Europe's space expenditure per capita is comparable to that of Japan. More than 90 percent of the budget available to ESA goes directly to European firms.[2]

Besides their ESA share, several members (notably France, Italy and Germany) have their own strong national space programs.   France has traditionally had the strongest commitments to space. This position must be seen in the context of its national defense policies, largely independent of the NATO partners.

Special emphasis is placed on European autonomy. For a long period, the primary aim of the German space activities was scientific research. One key area of activity was and remains bilateral space projects, principally with the United States. The Germans are reluctant to support the military use of space.

For Italy, the technological spin-off from space activities is seen as a justification for bigger investment than its gross national product would warrant (especially in telecommunications).   For the United Kingdom, satellite communications, partially for defense, have always been important, and over the last few years, the trend toward privatizing state activities has affected greatly UK space policy. The Scandinavian countries, Belgium and the Netherlands all hope that participating in European space programs will make their firms more competitive internationally.

ESA has its headquarters in Paris, with a staff of about 350 people. The technical center ESTEC in Noordwijk (Netherlands) employs a staff of 1,100. Its scientists and engineers guide and monitor project development work by companies, check proper functioning of satellites using ESTEC's own large test facilities, and carry out scientific and technological research.   The European Space Operations Centre (ESOC) in Darmstadt, Germany (300 staff members) controls a world-wide network of ground stations, and data received from satellites are collected and distributed to users. ESOC also looks after the satellites exploited by Eutelsat, Eumetsat and Inmarsat. ESRIN, the Information Centre located in near Rome, collects, archives, catalogues and distributes the data acquired by earth observation and scientific satellites and has 100 staff. ESA supports the French-operated Kourou launch base in French Guiana, where more than 200 engineers and technicians are employed.[3]

A major ESA activity has been to develop an independent launch capability. In the sixties, the first generation, called "Europe," was developed as a three stage rocket combining mainly British (Black Knight), French (Diamant and Veronique), and German (third stage cryogenic propulsion) rocket technologies. Since flight tests of Europe I and II were a failure, the program was terminated in 1973 and replaced by the Ariane development, which was conducted by ESA via the French space agency CNES, with Aerospatiale assuming overall industrial control. Although the complete rocket was flight tested in Kourou, individual launcher components are developed and tested in Europe by the involved companies (Aerospatiale, SEP, MBB/ERNO, BPD). Since its first successful launch on Christmas 1979, Ariane has conquered a growing fraction of the international commercial launch market (about 60 percent). By April 1992, 50

Ariane rockets had been launched, with a reliability of 96 percent in the last 25 launches. A private company, Arianespace, markets Ariane and manages the launches. With six different versions of Ariane 4--mixing pairs of solid and/or liquid/cryogenic-fuelled boosters--ESA can place satellites into low and geostationary orbits from Kourou.

With its Ariane-rocket, Europe profits from the internationalization of spaceflight. Most customers are from Europe or multinational organizations like Intelsat or Inmarsat. Therefore, it is in the interest of Europe to bring as many users as possible into the international space business. On the other hand, Arianespace is vulnerable when pitted against new competitors in the space launch market like China and Russia which can offer cheaper launch services. Therefore, Europe's interest in international space cooperation is limited to space application programs with those other countries which do not develop their own space launcher.

Even before they had a launcher, European countries had successful space programs (about 30 application satellites), initially using sounding rockets or launch capabilities from other countries. Activities extend to almost all civilian space application programs, especially in the area of scientific, Earth observation, and telecommunications satellites. With its SPOT-satellite, France accessed and commercialized moderately accurate satellite images. The Earth-Remote-Sensing satellite (ERS-1), carrying a synthetic aperture radar, provides Europe with radar satellite data day and night, independent of weather. The European Communication Satellite (ECS), launched in 1983 and later known as Eutelsat, was the first generation of communications satellites developed by ESA. The Maritime European Communications Satellite (MARECS) series was first launched in 1981. ESA's geostationary weather satellites (Meteosat) and communications satellites are operated by quasi-governmental organizations (Eumetsat and Eutelsat).

ESA conducts a broad program of space science and applications missions, which has stimulated international cooperation. Missions included Giotto, a spacecraft sent to Halley's Comet in 1985; Ulysses, launched in 1990 by the United States to explore the solar system; and Hipparchos, an astronomical observatory launched in 1989. ESA also contributed one of the main instruments for NASA's Hubble Space Telescope, and is building the Infrared Space Observatory. ESA is now participating with the United States in the Cassini mission to send a two-part spacecraft to Saturn in 1996. ESA's long-term space science program, called "Horizon 2000," includes space astronomy in different regions of the electromagnetic spectrum.

ESA has conducted some microgravity materials processing experiments (Eureca, Spacelab). With the Spacelab, developed largely by Germany, and launched first with the US Space Shuttle in the early 1980s, ESA gathered experience with manned orbital stations. Because of European dependence on

US space launch capability, the United States had an upper hand in negotiations with ESA. Consequently, Spacelab could only fly once under ESA control. Therefore, Spacelab was a turning point and strengthened European incentives for complete autonomy in manned spaceflight.

At their meetings in Rome January 30-31, 1985 and in The Hague on November 9-10, 1987, the Ministers responsible for space affairs in the ESA Member States decided upon a long-term plan, based on three main concepts:

- autonomy through new and competitive launchers
- coherence thanks to a combination of programs that would provide Europe with independent access to low Earth orbits and ultimately lead to a complete, inhabited European in-orbit infrastructure;
- balance between science, applications and infrastructure programs.[4]

Four major elements were included in the long-term plan:

1. Europe wants to build a manned orbital complex, comprised of its Columbus elements, including a permanently manned space laboratory attached to America's planned Space Station Freedom, an autonomous man-tended free-flying (MTFF) laboratory, and an unmanned platform for earth observation and environmental research;
2. The Hermes spaceplane to provide Europe with independent access to the manned space station elements and a capability for return to earth;
3. A new heavy launcher, Ariane 5, with a liquid hydrogen/oxygen rocket engine, to place payloads of up to 18 tonnes in low earth orbit, including the manned Hermes space transporter;
4. Finally, two powerful data relay satellites (DRS) to meet data transmission requirements between Hermes, the Space Station, and the unmanned ESA satellites.

The operation of the elements of the in-orbit infrastructure requires a considerable expansion of already existing ground facilities and in some cases, the establishment of new ones. The total cost of the first three programs from 1987 to the year 2000 was originally estimated at \$47 billion. In addition to these programs, Germany and the United Kingdom have national development programs to replace expandable launch vehicles by hypersonic aerospace planes.

By the time of the next Ministerial Meeting in Munich in November 1991, the situation had profoundly changed. Economic growth had slowed and fewer resources were available for research and development. Concurrently, Europe underwent a major geo-political upheaval, including the reunification of Germany, changes in the political system in the countries of Eastern Europe, and the acceleration of European integration.

In several European countries--in particular, in Germany due to unification costs--the low cost-benefit ratio and the negligible economic impact of manned spaceflight was criticized at high political levels.[5] The German Physical Society, among other professional science organizations, opposed the presence of humans in space to perform scientific experiments. Given the low contribution of space to the Gross National Product (less than 1 percent) and the rather small workforce, industry support in Germany is weak. In a widely acknowledged study, the German Fraunhofer Institute used statistical analysis of patents to ascertain that space technology cannot be justified by extraordinary spin-offs to non-space sectors.[6] Spaceflight is a recipient rather than a producer of technology. And in a multi-volume report on the proposed Sänger space transportation system, the Bureau for Technology Assessment at the German Bundestag supported similar doubts on the direct or indirect technological or economic effects of huge manned space programs.[7]

Although anti-space sentiments are not the same in all European countries, ESA had to respond to all these factors. In Munich, the member states restructured the Long-Term Plan, whilst reaffirming the guidelines established in The Hague. One year later, at the Ministerial Meeting in Grenada in November 1992, the proposals for the future of Europe's space programs were built around the following concepts:

- continuation of the major scientific programs (Horizon 2000);
- preservation of European competitiveness in launch (reduced Ariane 5) and telecommunications (Artemis, DRS-1) systems;
- expansion of the programs for observation of the Earth and its environment (ERS-2, Envisat, Eumetsat);
- greater cooperation in crewed spaceflight.[8]

What ESA calls a "flexible long-term, stepped approach" is actually a retreat from the original goals "autonomy, coherence and balance" expressed in 1985 and 1987. Without a free-flying space station and a manned, reusable space launcher (both programs have been canceled),[9] Europe cannot become an autonomous space power, at least in manned spaceflight. The new position takes into account the changing world environment and opens the way for a stronger involvement in international space cooperation.

Space Cooperation with the CIS

At the 1991 conference of ministers in Munich, ESA decided to perform a study on international cooperation in space, aiming at diminishing the financial burden and integrating additional know-how from other countries. The end of the Cold War and the opening of the former Soviet market presented new possibilities for cooperation with a space program, especially to develop new propulsion systems. Therefore, parallel to negotiations with NASA on Europe's involvement in the Freedom space station (Automated Transfer Vehicle, Assured Crew Return Vehicle, DRS-1), ESA negotiated cooperation with the former Soviet Republics in manned and unmanned spaceflight. In the field of space transportation, the Hermes program is to be re-oriented toward greater and deeper cooperation with Russia. An ESA-Russian program is being defined to include a crew transport system and servicing elements for low Earth orbits. With respect to the in-orbit infrastructure, definition studies are to be undertaken for a future ESA-Russian Space Station, including a contribution to the Mir-2 program.[10]

A comprehensive German study, called KOGUS, examined in detail the cooperation potential between Germany/ESA and the Commonwealth of Independent States (CIS) in spaceflight.[11] One outcome is a program called EUROGUS. KOGUS's key conclusions were:

- The space infrastructure of the CIS seems adequate for Western users;
- In most cases, potential CIS technical capabilities are far beyond West European needs; and
- In the area of space transport, the CIS has the capability to make any Western developments superfluous.

By cooperation with the CIS virtually every European space launch activity could be replaced, including Ariane. To diminish this "danger," European space industries and politicians want to control the CIS capabilities for development, testing and production of space launchers as quickly and cheaply as possible. This approach also offers the opportunity to catch up with the former superpowers in all relevant fields of launcher technology, including ballistic missiles.

Dual-Use and Vertical Proliferation of Military Space Technology in Europe

For the two superpowers, superiority in space was seen as the way to exert power against each other and against other countries. One hidden objective of European space programs was the desire of some elite factions to weaken the dominance of the superpowers via access to space technologies. This motive seems obvious for France and the United Kingdom, which as nuclear weapons states were involved in ballistic missile development. But in Germany, once the leading missile power, civilian spaceflight was also seen as a means to regain the lost technological leadership. As for missile proliferators in developing countries today, the inherent dual-use of space technologies provided the technical basis.

Missile Development

With its V-2 ballistic missile, Germany attacked European capitols in World War II and became the leading missile power in the world. Whereas Germany was not allowed to develop or produce its own missiles by the occupation agreements after the war, other countries exploited the know-how of rocket scientists from the V-2 team, including the United States, the Soviet Union, France, the United Kingdom, Argentine, Egypt, India, Brazil, Libya and Iraq.[12]

France had tight connections to the German rocket program until 1933. After the war, it hired several rocket engineers, including Eugen Sänger (the "father" of the orbital bomber) and his wife. About 70 experts from Peenemünde, the V-2 testing ground, developed the small Veronique and other sounding rockets, which provided the technical basis for the French military solid-fuel short-range ballistic missiles.[13] In 1958, Charles de Gaulle launched the production of land-based ballistic missiles with a range of about 2,500 km and the development of submarine ballistic missiles, which formed the nucleus of the force de frappe. In 1956, most German rocket specialists returned to Germany, but a few kept on working with the French launcher Diamant-A (Emeraude), the first stage of which was based on the V-2. This rocket was used to launch the first French satellite Asterix in 1965. After the failure of Europe-I, France constructed the Ariane out of Diamant-B and parts of Europe-II.

At the end of World War II, the United Kingdom had almost no experience in missile development, and hired about 20 German rocket engineers, among them the propulsion expert Walter Riedel. Their know-how was probably involved in developing the military liquid-fuel rocket Blue Streak and the three-stage intermediate-range missile Black Arrow, which looks similar to the first stage of the V-2. In 1971, the United Kingdom launched the satellite Prospero into space with this rocket. After cancelling the Blue Streak as a national program for nuclear delivery in 1957, the United Kingdom stopped national rocket programs and relied on US missile programs (Poseidon, Trident) and European space launch developments (Ariane). In the 1980s, British Aerospace and Rolls Royce began to develop a single-stage, reusable launcher called HOTOL (Horizontal Take-Off and Landing) with an air-breathing rocket motor.

These two examples demonstrate the close relationship and interchangeability between ballistic missiles and space launchers. Civilian programs generally followed the effort to develop a military missile.[14] One can also see how important a well-trained group of technical experts can be to initiate missile programs.

After 1955, West Germany became a leading power in all aspects of rocket technology, mostly in the context of space flight, but in military programs as well. Lacking the capability to test or produce long-range missiles, German engineers used other countries as testing grounds; and relied heavily on returning rocket experts (led by Eugen Sänger).

Today, the technical and organizational capabilities to be a developed rocket power can be found in organizations such as the German Research Establishment for Air- and Spaceflight (Deutsche Forschungsanstalt für Luft- und Raumfahrt, DLR, former DFVLR), the German Spaceflight Agency (Deutsche Agentur für Raumfahrtangelegenheiten, DARA) and the company Deutsche Aerospace (DASA). In many areas of missile and rocket development, German companies and research institutes are among the world leaders. Currently, great efforts are directed into the development of the Sänger spaceplane.

Of great importance is the German/French cooperation in the development of missiles with very short range, which is realized in cooperation between MBB and Euromissile. The most well-known and widely sold are the anti-tank and anti-aircraft missiles Hot, Roland and Milan. More advanced guided missiles projects are under research, development and testing. Experience with these programs, however, has little utility for long-range ballistic missile development.

Military Applications of Space Technology

By its constitution, the European Space Agency is obliged to engage in space cooperation for "exclusively peaceful purposes." In the early eighties, however, inspired by US President Ronald Reagan's military space program, some politicians in Europe began to see space as a matter of power, in military as well as in political terms. One official starting point was a speech held by the French President Francois Mitterand on February 7, 1984, in Den Haag, in which he argued:

> If Europe were able to launch its own manned space station allowing it to observe, transmit and consequently avert all possible threats, it would have taken a big step towards its own defence...To my mind, a European space community would be the response best adapted to the military realities of tomorrow.[15]

The Western European Union (WEU) undertook to host such a European space community. In a 1984 report WEU provided a comprehensive program to militarize European spaceflight. In a draft recommendation the WEU asserted that space capability will be a key determinant in future warfare, that in military terms the difference in potential between the space-capable nations and the others will be almost as great as the current difference in power between nuclear and non-nuclear nations.[16]

Besides military applications of missile and satellite programs, WEU discussed the direct involvement in weaponization of space, by developing anti-satellite systems and missile defenses. They say dual-use as a key to using ESA's capacities for military purposes.

Such a space strategy has been further developed and adapted to political changes by the German Society for Foreign Policy (Deutsche Gesellschaft für Auswärtige Politik, DGAP) in Bonn. In several studies, the DGAP demanded that satellites be used for reconnaissance, communications, early warning and warfare.[17] The 1990 study on reconnaissance satellites referred to disarmament verification and the surveillance of new threats emanating from the south as possible space-based security applications. A similar perspective can be found in a recent WEU study on European reconnaissance and verification satellites.[18]

As part of such concepts, France is expected to provide its SPOT satellites and an improved satellite called Helios for reconnaissance. Germany could modernize its ERS-1 radar satellite. For data transmission, the studies recommended the data relay satellite DRS. Data gathered from this system could be used for targeting in warfare.

Other European civilian satellite capabilities have implications for international security as well. Some civilian communication satellites perform military or military-related assignments or carry military components on board. The French Telecom I and II, for example, carries the Syracuse I and II, which are military payloads. Similarly, the Sirio satellite has been used by the Italian navy for mobile communications. Sicral, an Italian multipurpose satellite has served for military, national public security and civil protection.

A 1992 DGAP study proposed that the Sänger aerospace-plane be used to protect satellites and to transport military payloads into and out of orbit. A special demand for transport is recognized by the possibility of cooperating with the American SDI/GPALS or Star Wars program.[19]

Europe's Role in Space Trade and Missile Proliferation

The export market for space technology is still in its infancy, but it is expected to grow. The total size of the current export market for relevant technologies, goods and services can be only estimated. In the mid-eighties, the German Aerospace Industries Association (BDLI) projected for the 1990s a turnover in satellite and ground-segment exports in the order of one billion DM

($600 million). According to a EUROSPACE study of 1985, the annual space export of Western European countries would be $115 million for the years 1980 to 1984 and until 1990 $215 million per year.[20] These numbers are not very large. The world market continues to be dominated by US companies. Europe, however, is increasing its share in several sectors (space launch, telecommunication, and remote sensing). Governments are undertaking a variety of measures to help and to encourage firms to increase their share in the market of space technology, leading to mutual reproaches of protectionism.

Another political conflict has arisen between commercial interests and the potential dual use of exported space technologies for civilian and military purposes. The Missile Technology Control Regime (MTCR) has already affected the export of space technology and equipment in Europe. Before the MTCR came into effect in 1987, and even after, companies from Western Europe were among the leading suppliers of missile technology to proliferators in developing countries. This trend can be seen in the case of Iraq.

As the UN Special Commission on Iraq (UNSCOM) has proven, technology transfer and assistance from European and American companies and individuals have played a major role in building Iraq's indigenous capabilities from dual-use technologies. Most important was technology flow from Argentina's Condor II project to Iraq. In the late 1980s, for example, Iraq received German-made metal presses and other equipment to produce gas bottles, lampposts and milk separators. Actually, Iraq needed these metal presses to produce missile combustion chambers. In another case, a shipment of gyroscopes, motors and other missile components, which were listed for oil exploration by the exporting German company, were probably destined for missile launcher applications. According to UN documents and a German government report, sixteen German firms contracted to supply equipment used in Iraq's ballistic missile program, including turbopumps, rocket motor nozzles, high-pressure air intake systems, special welding components, high quality steel rods and pipes, and fuel systems.[21]

Some of the firms supplied Iraq with the machinery to make missile combustion chambers and fuel injectors. Others helped to redesign and to manufacture Iraq's missile gyroscopes, supplied plans for a complete fuel storage facility and helped to create and to equip a complete missile quality assurance program. Several of the companies are involved in space programs as well, as are MBB/Transtechnica, Aviatest or Mannesmann. Most of the companies produce dual-use goods in different fields, including testing and production facilities which can be used for a variety of purposes, including space and missile developments.

Other critical cases are the space launch programs of India and Brazil which strongly relied on cooperation and trade with Western Europe, in particular with France, United Kingdom and West Germany. India acquired knowledge from space cooperation in all relevant technology areas. A cooperation agreement signed between the German DFVLR and the Indian Space Research Organization (ISRO) in 1973 covered a wide range of activities, including training and educating Indian scientists on rocket propulsion; guidance technology; flight control; satellite positioning; remote sensing; materials science; shipping German equipment and software for sounding rocket programs; testing an SLV-3 model in a hypersonic wind tunnel of the DFVLR; constructing rocket test ranges; and studying reentry problems for delivery vehicles up to 1,000kg.[22] Indian scientists became familiar with composition, manufacturing, quality control, and error detection of composite materials (for example, glass fiber reinforced plastics), usable for production of heat-resistant rocket nozzles and missile warheads. Substantial support was provided in guidance systems, for example, by use of a German interferometer on an Indian sounding rocket in 1978, and by the Indian-German autonomous payload control rocket experiment (APC-REX) until 1989, which would provide autonomous (closed-loop), real-time navigation capability.[23] Furthermore, India gathered know-how for building a filament-winding machine.

On several occasions, MTCR restrictions and Western sanctions obstructed Indian-European space cooperation. In late 1989, the national implementation of MTCR regulations prevented France from selling cryogenic rocket engine technology to India. In 1992, the US State Department requested that Arianespace suspend its launch of India's Insat-2A satellite in view of US sanctions against ISRO.[24]

Foreign assistance and space cooperation from the beginning were important also in Brazil's missile development. With support from European countries, Canada and the United States, Brazil has developed four generations of solid-fuel sounding rockets (SONDA I - IV), which its companies Orbita and Avibras converted to military applications. Based on the SONDA series, Brazil designed a four-stage solid-fuel booster (the VLS).

Indigenous capabilities were developed also in a joint cooperation with Germany, agreed to in 1969. After 1973, the DLR organized workshops with Brazilian scientists on the whole range of missile-relevant topics.[25] Practical experience was gathered by joint rocket launches and the regular exchange of scientists. Germany provided direct technical assistance in rocket guidance, payload integration, and thrust vector control. A highly important topic was the development and building of a filament-winding machine, which has enabled Brazil since 1977 to manufacture rocket motor casings and thrust nozzles. The cooperation was extended to carbon-fiber technology, quality control, nondestructive testing, and production of rotor blades. Systems integration and vibrations testing of a SONDA-IV stage were performed at environmental testing laboratories of the German aerospace company MBB.[26] Brazil frequently resold know-how with military applications acquired from Western nations for civilian purposes (for example, to Iraq). Recently, the Ariane consortium planned to assist Brazil in building a rocket engine, which led to US opposition.

European Attitudes Towards Missile Proliferation:
Between Missile Defense and Missile Control

The Gulf War marked the decisive event in Europe's attitude towards missile proliferation. The European public and politicians perceived (partially inspired by press reports) that in future wars their own capitols, like Tel Aviv, might become targets of "mad dictators" attacks. However, hysteria vanished after the war, as it became clear that in the next few years, no developing country would be either technically able or politically willing to threaten Europe. People also were less frightened by the poor man's weapons of mass destruction than they were by the gigantic overkill capacities of the former superpowers.

European Missile Defense Initiatives

But certain military circles, bereft of missions after the collapse of the Soviet Union, saw proliferation as a way to maintain and redirect military arsenals. Inspired by George Bush's GPALS initiative, the concept arose of a European Protection Against Limited Strikes (EPALS), promoted by High Frontier Europe, the counterpart of the conservative American High Frontier organization which stimulated Reagan to launch Star Wars. These ideas began to circulate on governmental levels as well. In November 1992, the WEU took the initiative by preparing a report on anti-ballistic missile defense.[27]

In a draft recommendation, being part of this report, the Assembly recognized that Europe is "no longer threatened by a ballistic missile attack from the territory of the former Soviet Union," but recalled that "the danger of proliferation of ballistic technology and nuclear, biological and chemical warheads stockpiled on the territory of the Commonwealth of Independent States has not yet been averted." Noting that "several third world countries, particularly in the Mediterranean and the Near and Middle East, are making considerable efforts to procure ballistic systems capable of reaching European countries," the Assembly recommended that risks of missile proliferation be assessed in order to facilitate a joint European position towards GPALS. It also instructed its Technological and Aerospace Committee to pursue its work on anti-ballistic missile defence problems and to organize a symposium in 1993 on this subject.

The WEU report surveyed recent European activities in the field of missile defense. Several SDI memoranda of understanding (MOU) had been agreed upon between the United States and the United Kingdom, Germany, Italy, Israel and Japan. The United States financed work worth $833 million by industries in about ten countries.[28] In seven years, Germany received less than $100 million.

According to the WEU report, industries have studied the feasibility of a ballistic missile defense architecture in several Western European countries, either at the request of government or on their own initiative. These initiatives include the agreement concluded in 1986 between the French firms Aerospatiale and Thomson-CSF whereby they set up the economic interest group CoSyDe (defense systems concepts) to develop weapons systems capable of countering the threat of ballistic or non-ballistic missiles in the European theater.

Low-altitude interception functions comparable to Patriot and Erint missiles are currently being developed in Europe. The Franco-Italian consortium Eurosam (Aerospatiale, Thomson-CSF and Alenia) is now developing the Aster/ARABEL anti-aircraft missile with anti-missile interception capability. This group is also working on the development of two ground-to-air missile systems, one of which, the SAMP-T (surface-to-air medium-range) missile, might be given a limited anti-ballistic missile capability. Another enlarged air defence concept is being considered that would establish a limited anti-theater ballistic missile defence capability based on the tactical air defence system TLVS (Taktisches Luftverteidigungs System).

In regard to high-altitude interception, a number of European industries, including the French firms Aerospatiale and Thomson-CSF, are sub-contractors in the development of the American THAAD (theatre high-altitude defence) system, designed to protect areas of more than 10,000 km2.

The German government's position on anti-missile defense systems has not been decided. The prevailing opinion in the Ministry of Defense is that Germany is not threatened; but it recognizes that a potential threat may exist along the southern flank of the alliance. The tendency, therefore, is to tackle all matters relating to ballistic threats in the framework of NATO.

European Non-Proliferation Regimes

The WEU report suggests that European countries need to address many questions:

It is therefore time to hold a public debate to determine what Europe needs to guarantee its future security. The debate should concentrate on two main topics: the possibilities of building on international law as an instrument of security on the one hand and the measures necessary for improving defence and protection arrangements on the other. However, it should be underlined that these two approaches cannot be alternatives. They are complementary.[29]

The WEU report envisages the following steps towards a non-proliferation regime which represent a typical European position today:

Legally, every effort should be made to improve and extend the nuclear non-proliferation regime with a view to making it truly universal including chemical and biological weapons. In particular, work should at last be completed on a universal convention banning the production and dissemination of chemical weapons.

Furthermore, steps should be taken to perfect and extend the missile technology control regime (MTCR) on which most encouraging progress has already been made. With particular regard to space, the question is how far will it be possible to build on international law to set limits for the military use of space. It should be recalled that the January 1967 space treaty banned the placing in orbit of warheads of mass destruction and the placing of military installations on celestial bodies, but made no provision for the demilitarization of outer space as such. Conversely, the draft treaty submitted to the United Nations General Assembly in 1981 provided for a ban on the emplacement of weapons of all kinds in outer space.[30]

It is important to note that WEU discusses the two questions of missile non-proliferation and demilitarization of space in conjunction. Since the space weapons issue was addressed earlier, I will discuss only the MTCR, with a special focus on Germany.

Recent export scandals (for example, the Rabta case or the Saddam experience) have led to public discussions about the effectiveness of European export controls since 1989. Germany's export control was based on a liberal economic policy and the principle that exports were unrestricted, except in cases in which exports may undermine national security interests. Therefore, in the Iraq case many of these exports were not illegal as they were not subject to control or they were based on false end use declarations or were approved as dual-use commodities. As a result, the German Regulations on Foreign Economy (Aussenwirtschaftsverordnung, AWV) and the Export Commodity List (Ausfuhrliste, AL) have been revised several times and a new enhanced structure has been created.[31]

In January 1992, the German Bundestag passed a draft amendment to the Federal Trade Law to prevent illegal arms trade. The amendments allow for prison sentences of up to five years and the imposition of large fines, and call for the full confiscation of profits made from illegal arms trade. Existing laws already authorize the monitoring of postal and telephone communications. The department responsible for export control will be separated from the Federal Office for the Economy. To control the national implementation of the MTCR, a special technical group for MTCR tasks handles the export controls and decides on the technical aspects of dual-use items, while representatives from ministries decide on the political relevance.

The European Commission is attempting to harmonize national export controls on dual-use products and technology, including missile technology. Germany is especially interested in the initiative because it has the strictest export control law and German companies might suffer from national differences. Unless other nations harmonize their controls to be as stringent as German standards, the reform might be reversed.

Conclusions

In general, European countries are in favor of avoiding new missile threats which could reach European territory or might otherwise lead to international instability. Currently, emerging missile threats from developing countries are not a high political priority in European politics, but this may change if European capitols come within flight distance of missiles from developing countries.

In Europe, the MTCR is perceived as being effective in slowing down missile programs in developing countries and focusing political support on missile non-proliferation. Of the 22 members in mid-1993, 18 are from Europe. However, the limits of the MTCR are recognized, especially its discriminatory character and the conflict with commercial space programs that prevents certain cooperative space programs with developing countries. Therefore, most European countries would probably favor demand-side oriented solutions, based on missile arms control and disarmament which would not block Europe's space efforts.

The European space power would profit from any attempts to internationalize spaceflight, as long as its competitiveness in the commercial launch service is not undermined too much by emerging launch capabilities. Europe can offer launch service and a wide range of space application technologies to developing countries.

Due to the poor economic situation in Europe and the cuts in the space program, European leaders and policy makers are skeptical of unnecessary restrictions and additional costs of a non-proliferation regime, but might finally accept it if convinced that the security gain was significant.

To achieve worldwide limitations on ballistic missile technology, the effectiveness and intrusiveness of the verification mechanism is decisive.

Inspections of space facilities, which are seen as too intrusive and costly, would diminish European readiness to accept a control regime.

France and the United Kingdom might have strong reservations about giving up their nuclear missile force as part of a non-proliferation regime. They would have to be convinced that the missile arsenals of the former superpowers or of developing countries have been reduced or eliminated before they would go so far in this direction.

European missile defense programs have diverted the debate into a conceptual cul-de-sac. If implemented, these programs might diminish European interest in missile non-proliferation policies. The future of Star Wars in the United States has some, but not a decisive impact on European efforts to build a ground-based defense using Patriot-type technology. Deep reductions or the complete elimination of ballistic missiles, including those from North and South, would undermine European incentives to build a ballistic missile defense.

Neither of the former superpowers seems interested in space weapons. There is a window of opportunity to negotiate an international ban on space weapons. European governments, including France, have expressed several times their interest in such a proposal.

Notes

1. The members are Austria, Belgium, Denmark, France, Germany, Ireland, Italy, Netherlands, Norway, Portugal, Spain, Sweden, Switzerland, and the United Kingdom. Finland is an associate member and Canada has signed cooperation agreements.

2. R. Lüst, "European Space Programmes," in Space Course, Aachen 1991, p. 12.

3. Lüst, "European Space Programmes," p. 12.

4. J.M. Luton, "The Granada Ministerial Conference--The Issues and the Outcome," esa bulletin, n. 72, November 1992, pp. 8-11.

5. On the German debate see W.M. Catenhusen, W. Fricke, eds., Raumfahrt Kontrovers, Bonn: Forum Humane Technikgestaltung, Heft 3, 1991; J. Weyer, ed., *Technische Visionen--Politische Kompromisse,* Geschichte und Perspektiven der deutschen Raumfahrt, Berlin: edition sigma, 1993.

6. U. Schmoch, N. Korsch, Analyse der Raumfahrtforschung und ihrer technischen Ausstrahlungseffeckte mit Hilfe von Patentindikatoren, Karlsruhe: Fraunhofer-Institut für Systemtechnik und Innovationsforschung (ISI), 1990.

7. H. Paschen, R. Coenen, F. Gloede, G. Sardemann, H. Tangen, Technikfolgenabschätzung zum Raumtransportsystem "SÄNGER," Bonn: TAB, June 1992. The study proposed a political choice between a "conservative" space scenario, limited to unmanned space application programs, and the "progressive" scenario, including manned missions to Moon or Mars. Only the latter might justify the large development costs for Sänger.

8. Luton, "The Granada Ministerial Conference."

9. Hermes was reduced to a technology demonstration program, with a three-year re-orientation phase. Its costs and weight have grown; it is too heavy for the Ariane 5 now under development; a new version designated Ariane Plus, was proposed.

10. M.K.E. Hauger, Deutsch-russiche Kooperation in der Raumfahrt, Luft- und Raumfahrt, April 1992, pp. 12-20.

11. KOGUS, Kooperation von Deutschland/ESA mit der Gemeinschaft Unabhängiger Staaten in Infrastrukturbereich der Raumfahrt, Summary/Final Report of the Weltraum-Instituts Berlin for DARA, October 13, 1992. GUS is the German synonym for CIS.

12. For details see J. Scheffran, "Die heimliche Raketenmacht: Deutsche Beiträge zur Ausbreitung und Entwicklung der Raketentechnik," Informationsdienst Wissenschaft und Frieden, January 1991.

13. These French sounding rockets later became the technical basis for Pakistan's Hatf missiles. See S. Chandrashekar, "An Assessment of Pakistan's Missile Capability," Missile Monitor, no. 3, Spring 1993, 4-11.

14. See J. Scheffran, "Dual-Use of Missile and Space Technologies," in G. Neuneck, O. Ischbeck, eds., Missile Proliferation, Missile Defense and Arms Control, Baden-Baden: Nomos, 1993.

15 "The Military Use of Space," Committee on Scientific, Technological and Aerospace Questions, WEU Document 976, May 15, 1984, p. 197.

16 Ibid., p. 183.

17. See Deutsche Weltraumpolitik an der Jahrhundertschwelle, Report of an expert group, Bonn: DGAP, 1986; Europas Zukunft im Weltraum, a common report of European institutes, Bonn: Europa Union Verlag, 1988; Beobachtungssatelliten für Europa, report of an expert group, Bonn: DGAP, 1990.

18. Assembly of the Western European Union, Official Record of the Symposium "Observation Satellites: A European Means of Verifying Disarmament," Rome, March 27-28, 1990. Paris: WEU.

19. K. Kaiser, Außen- und sicherheitspolitische Aspekte des Raumtransportsystems SÄNGER, Research Institute of the DGAP, January 1992, p. 19.

20. S.F. von Welck, "The export of space technology: Prospects and dangers," Space Policy, August 1987, pp. 221-231.

21. The Washington Post, July 23, 1992. More details can be found in: K.R. Timmerman, *The Death Lobby: How the West Armed Iraq*, New York: Houghton Mifflin, 1991.

22. R. Rudert, K. Schichl, S. Seeger, Atomraketen als Entwicklungshilfe, Marburg 1985, p. 99.

23. G. Milhollin, "India's Missiles--With a Little Help from Our Friends," Bulletin of the Atomic Scientists, November 1989, pp. 31-35.

24. This information was found in the database of the International Missile Proliferation Project in Monterey.

25. Rudert, et al., "India's Missiles," p. 56.

26. M. Birkholz, et al., Die Bundesrepublik als heimlicher Waffenexporteur, Berlin: Arbeitskreis Physik und Rüstung, 1983.

27. Report on "Anti-ballistic missile defense," Assembly of the Western European Union, 38th Session, Document 1339, November 6, 1992.

28. WEU Report, 1992, p. 13.

29. Ibid., p. 15.

30. Ibid.

31. J. Scheffran and A. Karp, "The National Implementation of the Missile Technology Control Regime: The US and German Experiences, in H.G. Brauch, et. al., *Controlling the Development and Spread of Missile Technology*, Amsterdam 1992, pp. 235-255.

6

International Space Cooperation and a Non-Proliferation Regime: Turning Plowshares into Swords?

Joan Johnson-Freese

In 1957 Wernher von Braun watched in frustration as repeated attempts to launch the first American payload into space ended in flames on the Cape Canaveral launch pad. The Navy Vanguard rocket, he knew, was not as technologically mature as the Redstone that his Army-directed team had under development. What he did not know was that capability was not driving the government's approval of Vanguard over Redstone as the vehicle to place the first US satellite into orbit. Rather, establishing the legality of satellite overflight was the primary government goal, giving projects with strong civilian flavor an important edge.[1] The Vanguard was primarily a scientific rocket whereas the Redstone was intended to carry a tactical nuclear weapon. Only after the public outcry that followed Sputnik was von Braun finally allowed his chance, and he successfully launched the Explorer I satellite from a modified Redstone missile. With a change of payload, the military missile had become a space rocket with a civilian payload. A sword had become a plowshare.

Today, the United States and the former Soviet Union are greatly reducing their missile arsenals. At the same time, however, countries are undertaking civilian space activities to gain practical benefits in areas such as communications, remote sensing, and meteorology. Often, these activities are conducted on a cooperative basis, to share costs, maximize data returns, increase technical expertise, or a variety of other reasons. The question arises whether increased cooperative civilian space activity could impair nuclear non-proliferation efforts through the spread of the most threatening delivery system, the rocket/missile.

My purpose in this paper is to examine the links between civilian space activities and nuclear non-proliferation efforts. Specifically, I inquire how increased cooperative space activity might affect non-proliferation efforts, if at all. Also I consider whether some formal regime or even a World Space Agency (WSA) for controlling the technology would be either desirable or feasible, based on past experience. It is my thesis that increased space cooperation does not increase the risk of missile technology spreading, largely because rocket technology is not normally acquired cooperatively. Rather, developing rocket technology has been highly nationalistic. Moreover, I argue that cooperative space activities do not rely on institutionalized cooperation to exist. Nor should they.

The Non-Proliferation Regime

The Nuclear Non-Proliferation Treaty (NPT) has been the primary international agreement to contain nuclear proliferation since 1968. The United States and the Soviet Union promoted the NPT in their pursuit of perceived national interests. The Treaty divided the world into two categories of countries, nuclear-weapon-states (NWS) and non-nuclear-weapon-states (NNWS), and imposed different obligations and constraints on each. Many of the countries that subsequently signed the Treaty, particularly the early signatories, did so because they too believed it would serve their interests. Article IV of the Treaty "rewarded" countries that surrendered their right to acquire nuclear weapons with access to nuclear energy technology. Those rewards, however, were slow to materialize, especially after the passage of the Nuclear Non-Proliferation Act (NNPA) of 1978 in the United States.

The NNPA was aimed to establish new, more stringent, non-proliferation controls to complement the NPT. But outside the United States the NNPA was seen as a unilateral imposition of a US-defined "consensus." Non-nuclear-weapons states repeatedly charged that the NNPA violated Article IV. Ironically, the NNPA was also criticized for undermining non-proliferation efforts by driving countries to develop fuel cycle self-sufficiency.[2] A number of countries did not sign the NPT, including Argentina, Brazil, India, Israel, and Pakistan. They continued with their nuclear plans, causing many NPT signatories to question the wisdom of their own commitment.

The International Atomic Energy Agency (IAEA) is charged under the NPT to promote nuclear power. The Agency also has the concurrent responsibility to restrict the production and transfer of nuclear fissile material. There have always been concerns about its effectiveness. As Walter Patterson wrote:

Such restriction, policing the world's breeding grounds for nuclear weapons, was conceived as a main responsibility of the International Atomic Energy Agency, from its inception in 1956. But its success was then and remained limited. In March 1962 the Agency safeguards system came into being. An incidental difficulty was that it could only be exercised when a national government permitted it to be--a very few national governments were so inclined. It is hard to feel any surge of confidence about the essential efficacy of the present safeguards arrangements; and in fairness it must be said this unease is shared by many Agency safeguards staff members.[3]

Several characteristics of the NPT/IAEA safeguards system should be noted when it is evaluated as a potential model for other prospective "control" or "monitoring" organizations. First, compliance with IAEA requests for on-site inspections (except for the defeated Iraq) is strictly voluntary on the part of the member states. Second, the IAEA has no real enforcement capability (except in Iraq). The Agency relies largely on the power of public censure to persuade a country to alter its behavior. Third, the IAEA has become highly politicized. During one incident in 1982, for example, the US delegation walked out of an IAEA General Conference to protest against its withdrawal of Israel's credentials.[4]

Moreover, countries with sophisticated technological capabilities are capable of producing nuclear weapons, and could have the parts available and ready for assembly without violating any international guidelines. National security interests could dictate the wisdom of such an approach, particularly in regional "trigger" situations. For example, if North Korea were to produce a nuclear weapon, Japan might feel compelled to reciprocate, which could then nudge Taiwan and/or South Korea to proliferate.

Since the NPT was signed, the international system has become more complex. The categories of nuclear-weapon-states and non-nuclear-weapon-states are inadequate to describe the proliferation dilemma. Some countries seek to acquire technology; some countries seek to sell technology; and some countries seek to prevent the sale of technology; sometimes the categories overlap. In addition, some countries are openly nuclear-armed; some are covertly nuclear-armed; some are "temporary" nuclear states; some are ambivalent proliferators; some are hard-core proliferators. No simple dichotomy reflects the fluid reality today.

The Missile Technology Control Regime (MTCR) has been the primary mechanism for restricting transfers of missiles and related technology. Initiated in 1987, the MTCR is not a treaty, but a set of voluntary export guidelines that each of the twenty-four member countries (plus three which have declared their adherence to the guidelines) implements in accordance with its national legislation. Although judged a success by most analysts and supported by non-proliferation advocates, the MTCR is restrained by limitations inherent with any voluntary mechanism: it is often slow because of its consensual operating nature and agreements reached by the group sometimes reflect the minimally desirable position. But, politics being the art of the possible, it may be the most effective regime that is possible. Should the MTCR become a formal organization? Should a World Space Agency be created to promote, among other objectives, non-proliferation of missiles to complement the MTCR? I return to these two questions after reviewing past, present and prospective cooperation in space in the next section.

The Changing Nature of "Space"

The international political landscape is changing so fast that analysis is rather like narrating a baseball game. Recent events in Eastern Europe and the former republics of the Soviet Union, the rise of Japan as a technological leader, increased integration in Europe, international environmental concern, the growing US budget deficit, and global economic recession have all had far-reaching effects on the space community. Eastern Europe and the former Soviet Union, for example, could concentrate on their ailing economies and abandon their interest in space. In the United States, one cannot assume any "peace dividend" for investing in space from the change in political relations with the former Soviets while the national deficit continues to grow.

In the 1980s, the primary motivation for space-related activity of both the United States and the former Soviet Union was their political competition rooted in national security concerns. This competition often took the form of a "race," for various reasons. First, there was a *satellite race,* initially for military surveillance and later for the worldwide communication market. Second came the *man in space race* in search of prestige and technological prowess with which to lure other countries during the Cold War years. Third there was the *planetary race.* The scientific knowledge to be gained from reaching the planets was and is significant. But the goal was again prestige. And finally, in search of the ultimate prestige, the moon race was exemplified in the Apollo program.

After national security, a variety of driving factors impelled the superpowers to embark on these races. Science has often motivated space programs, and encompassed medicine, astronomy, meteorology, oceanography and basic physics. But basic science yields few immediate returns. Nor have space programs as yet contributed much to fundamental theory. Consequently, science in space has faced many lean years--or years of intense political competition for resources.

Space has also been viewed as an area for commerce. This activity has been limited by the high up-front investment required. The high cost (running at about $3,500 to $14,000 per-pound-to-orbit) and long payback period has inhibited innovative new commercial opportunities in space. Space can be used also to obtain remotely sensed data for use in development. Finally, exploration and expansion of spheres of human activity is the most nebulous of motivations, but one which cannot be ignored.

Relative political stability between East and West increases the opportunities for international cooperation in space in ways that yield tangible and useful benefits. Satellite disaster warning systems, use of remote systems for monitoring the environment, and even international exploration missions such as a Mars venture all embody the spirit of international political cooperation in a highly visible way. The latter imperative may have become a driving factor in space cooperation, and may in turn reinforce the impact of science, education, and technology development. Commercial returns from space, however, are determined mostly by economic competition and appear to be oblivious to the state of politics.

Past Cooperative Efforts

Even during the Cold War, there were cooperative space efforts between the United States and the Soviet Union concurrent with their arms agreements. Space science began with a bilateral agreement signed in 1962 by National Aeronautics and Space Administration (NASA) and the Soviet Academy of Sciences. The 1962 Dryden-Blagonravov agreement, so called because it was negotiated by NASA's Hugh Dryden and the Soviet Academy's Anatoly Blagonravov, stipulated that national efforts would be coordinated in the fields of meteorology, geomagnetism, and satellite communications experimentation. The results were disappointing, in part because of the then poor quality of Soviet data processing technology, but also due to the Soviet penchant for secrecy.

The space arms control treaties signed by the United States and the Soviet Union in the 1960s and 1970s do not in themselves argue well for the prospects of expanding arms control in space. Indeed, most of the treaties were public relations gestures. As Walter McDougall wrote:

Thus the UN Outer Space Treaty of 1967, ratified by a vote of 88 to 0 on April 25. It denuclearized outer space and demilitarized the moon. But it did not demilitarize outer space. As for space being "for the benefit of all peoples irrespective of the degree of their economic and scientific development," the negotiators described it as a vague principle with no foreseeable application. In terms of the "space for peace" globalists, therefore, the space treaty was all show and no substance.[5]

In the space-related arms control field, the United States and the Soviet Union could afford to give up a space arms race that they didn't really want anyway. They were equally content to establish a status quo in which they could retain a proven technology and deny the same to others by disallowing testing. It is interesting that the pertinence of the Treaty was not significant until the 1980s with the Reagan Administration's Strategic Defense Program. Then, the Soviets publicly held the United States to the ABM Treaty, much to the chagrin of some individuals in the Pentagon.

For example, the Anti-Ballistic Missile Treaty (ABM), negotiated as part of the SALT I Treaty in the early 1970s, was portrayed as the first step to reduce nuclear arms. Again, it was mostly show. As Nathan Goldman argued, "The ABMs would be an expensive new weapon anyway. The treaty thus gives up the expensive, untried system for the promise of future arms control."[6] The space-faring nations were far less cooperative as soon as they confronted forgoing something either considered valuable or potentially valuable. No major space power, for example, has ratified the Moon Treaty which limits or even prohibits space mining in accordance with under its "common heritage" principle.[7]

The most successful institutionalized cooperative ventures, such as the International Telecommunications Satellite Consortium (INTELSAT) and the International Maritime Satellite Organization (INMARSAT), are peaceful, purposeful, and profitable activities. The organizations were created during the formative years of the underlying technology. Other than these examples, there is little evidence of potential cooperation in space.

Space science is the area of space activity most suitable and generative of international cooperation. From the earliest days of civilian space activity, NASA has cited the "conduct of projects and activities having scientific validity and mutual interest"[8] as an official reason to engage in international cooperation. International collaboration has also been a principal way whereby many countries nurtured space programs along the technical learning curve.

Why has space science traditionally been regarded as an area amenable to international cooperation? There are two basic reasons. First, it is mostly beyond national security interests. Second, space science usually has been a low priority for funding within national space programs. Scientists have had to cooperate to maximize scientific returns given their limited resources. This latter search for synergy explains most of the success of space science cooperation.

In non-scientific fields of possible space cooperation, political and economic factors dominate. Cooperation may result in technology transfers that are contrary to commercial interests. Space science, therefore, is a unique domain of international cooperation from which little can be extrapolated to other fields of space activity.

The transfer of space transportation technology from the United States to Japan provides a good example of the axiom that "politics always wins"[9] when pitted against cooperation. Japan was able to buy rocket technology from the United States during a period where such sales were prohibited to the rest of the world. One US State Department official in the early 1970s was able to promote such a technology transfer agreement between the two countries in spite of the reservations of NASA and the Department of Defense (DOD). As far as NASA was concerned, transferring this technology simply did not fit into their criteria for cooperation.[10] DOD's concerns were focused on the symbiotic military-civilian nature of space technology. But the State Department was able to logroll the DOD and NASA by arguing that the agreement would improve political relations between Japan and the United States.

Because of the unique circumstances, the United States-Japan example was clearly an anomaly, not likely to be repeated. States are likely to continue and expand cooperation in areas like space science, and perhaps remote sensing for humanitarian purposes. Given the thin history of international cooperation in the non-proliferation and space fields, is it feasible to build a new regime restricting long range missile technology?

Space and Missile Non-proliferation

The end of the Cold War has revived the concept of a world with zero ballistic missiles (ZBM). Strategic analyst Alton Frye suggests that a ZBM would be beneficial because:

- Strategically, a world without ballistic missiles would be far more stable than one in which missile proliferation makes countries vulnerable to nearly instantaneous attack from many quarters.
- Technically, constraining ballistic missiles is more feasible than preventing the spread of clandestine nuclear capabilities.
- Politically, the goal of banning ballistic missiles promises a new consensus to heal the fractures within the American strategic community. Advanced as a truly non-discriminatory regime, ZBM should also attract even wider international support than the Nuclear Non-Proliferation Treaty, which enjoys nearly universal, albeit fragile, adherence.[11]

However, Frye also points out that it would be very difficult to distinguish between missile tests and space launches,[12] rendering problematic the monitoring of compliance with a missile ban. Others disagree with this premise. Clearly though, if a country supposedly executing a space launch were found to actually be testing a missile, it would be difficult after the fact to do anything about it.

Frye notes that most nations regard access to space for communications, remote sensing, and a variety of functions as desirable and even a prerequisite of a modern society:

> [A ZBM treaty must] deal with the reasonable demand of non-spacefaring countries to share the benefits of space programs. The Outer Space Treaty of 1967 anticipated the spirit of the NPT by stating that 'the exploration and use of outer space should be carried on for the benefit of all peoples irrespective of the degree of their economic or scientific involvement.' That commitment has largely endured, and countries in every corner of the planet have enjoyed the fruits of space programs in global communications, meteorology, environmental monitoring, and other fields.[13]

This idea of "sharing the benefits" has striking similarities to Article IV of the NPT, benefits which were not always forthcoming. What would a missile non-proliferation regime that promoted diffusion of space launch technology have to achieve to prevent the spread of missiles, given the symbiotic nature of rocket/missile technology? The question is similar to that which arises in the NPT concerning fissile material. First, space-aspiring nations would have to be assured of access to space. Second, nations selling space launch technology or services would have to earn a reasonable return. Third, on-site and possible intrusive inspections of space launch facilities and payloads may be required to ensure that facilities and technology were used only for peaceful purposes. The third point would be of particular relevance to the United States, where congressional approval is needed to ratify treaties.

Except for the sharing of US rocket technology with Japan, the brief history of space offers few examples from which one might extrapolate in relation to assured launch technology transfer. Dependence on US launch vehicles became a particularly acute issue in trans-Atlantic relations after Symphonie, the first European (Franco-German) communications satellite, ran into problems with the US government. Because it was an operational communications satellite, the US-dominated INTELSAT consortium regarded Symphonie as competition. Hence, the United States mandated that Symphonie could be used only for experimental purposes. That incident provided the French the excuse they had been waiting for to secure German support for development of an autonomous European launch capability. Today, even with a diverse commercial launcher market, launch autonomy remains an important consideration for some countries; because of the key role being postulated for space in the future concerning national security, and because some countries do not want to have to submit to payload scrutiny or approval.

The two basic trends in national space activity today are: (1) increased cooperation in some areas while living with competition in others, and (2) the search for ways to decrease dependence. Based on the past NPT experience, it is highly doubtful that space-aspiring countries will agree again to codify their dependence on other countries for a critical capability. In part, they will resist doing so because assured access to space has a military component. Nations want to be able to launch satellites for command, control and communication and for surveillance before, during and after wars. Suppliers of such technology, in turn, will be reluctant to endow adversaries with such capabilities, either directly or via third parties.

The major impulse toward missile proliferation remains the political ambitions of proliferating states exploiting the opportunities from a depressed commercial launcher market. It is equally obvious that a hard-core proliferator determined to obtain missile capabilities will do so, regardless of incentives to forgo or control this capability are put in place. The history of nuclear proliferation demonstrates this fact. States have acquired nuclear weapons by a variety of means, including civilian energy programs, clandestinely, and by circumventing or simply ignoring institutionalized mechanisms to prevent proliferation. The situation with missile technology is analogous.

Unfortunately from the perspective of a non-proliferation regime, the commercial launch field is currently a buyer's market. There are more launcher rocket families than can be supported by the number of annual launches. The market for commercial launches is approximately 20 per year. According to one authoritative analysis, a launcher family needs to lift into space about 12 satellites annually (which corresponds to about eight launchers) to achieve a sound commercial return.[14] With the Chinese Long March, Ariane-44L, US Titan III, Delta, and Atlas, and Russian Proton all commercially available already and the Japanese H-II (and J-I) soon to be available, the competition will be intense, even considering the different payload capabilities of the launchers.

Some launchers are more commercially viable than others. The Chinese Long March is equivalent approximately to US Delta vehicles in payload capacity, but has a much lower reliability rate. It is also priced below the Delta and offers a shorter "wait" time from order to launch. The market for launching satellites has opened to international bidding since 1990 and international competition is intense. The Japanese H-II has little time in which to demonstrate its reliability and low cost. It seems unlikely, therefore, that the H-II will be competitive internationally. A representative of prime contractor Mitsubishi Heavy Industries (MHI) was quoted in 1990 as saying, "If everything goes well, we would like to be like Europe's Arianespace, but that is only a dream now."[15] That dream has slipped further and further away. MHI Space Systems

Department Manager Hiroshi Saito said in 1992 that the H-II is probably not suited for commercial purposes, as it is simply too expensive.[16]

Drawing from the Japanese experience, and considering their acknowledged technical capabilities, there seems little economic rationale for any country to spend scarce resources on ballistic missile technology development to launch civilian satellites. There are many problems in the space transportation field (primary among them being the high cost-per-pound-to-orbit issue noted earlier), but lack of enough ballistic launchers to satisfy the market is not one of them. As launching is a commercial field involving many companies from several countries, options are plentiful and the risk of dependence on foreign launchers is primarily political. It follows that incurring the expense of developing an independent ballistic launcher system is likely to be prompted by political or military objectives. Conversely, although the over-saturation of the launcher supply market is an argument against developing new ballistic launchers, it also tempts sellers into imprudent but potentially lucrative sales. As Aaron Karp put it:

> Industrial nations, eager to subsidize their own programs, are always on the lookout for foreign capital. To beat the competition, they are increasingly willing to submit to third world demands for technology transfer. France, for example, has replaced the United States as the leading supplier of space-launch technology to the Third World, having concluded agreements with Brazil, India, Indonesia and Pakistan. The prospect grows continually that future headlines will announce a European sale of [space launch vehicles] SLVs to a third world client. Once this taboo is violated, other exporters will face the same pressures to do the same.[17]

Further, even if hardware is not available from industrialized space-faring nations, the services and know-how of foreign engineers often are accessible. Expertise from the former Soviet Union has been of particular recent concern in this regard.

Much like the experience of non-proliferation efforts between 1960 and 1980, some countries may prove to be unstoppable in their quest for ballistic missile technology, regardless of cooperative efforts to stop them. It should be remembered, however, that not all countries are "hard-core" proliferators; indeed there is a spectrum of intentions and situations ranging from the so-called "temporary nuclear states," like the Ukraine (although whether their nuclear capability will be only temporary is questionable), to ambivalent nuclear states, to those whose intentions are primarily dependent on the actions of other regional countries. Therefore, finding the most feasible approach to managing this dangerous trend becomes a critical issue, and institutionalizing international civilian space efforts is an obvious option to consider.

Institutionalizing Space Cooperation

I have argued that space-related cooperation has evolved into two basic types: that involving services-for-profit obtained in an open market (for example, INTELSAT and INMARSAT); and that involving non-competitive, "low-politics" areas (for example, space science and the environment). The Committee on Earth Observation Satellites (CEOS) is a good example of the latter. CEOS is an international group created to facilitate cooperation between public satellite system operators, and the developers of new sensors and new satellite techniques.[18] It illustrates the possibilities and limits of operating outside the market. The objectives of CEOS are clearly worthwhile and its institutional framework has facilitated international cooperation in Earth observation. But CEOS lacks any authority, and rests on non-enforceable commitments from participants to cooperate. Data users therefore have no guarantee that the information on which they rely will be available on a continuing basis. Indeed, it may be available only as long as the budget officials in the participating countries see fit to continue funding. Experiences with programs funded by only one government, such as LANDSAT and SPOT (France), show that government funding may lapse with little warning.

The Inter-Agency Consultative Group (IACG) for Space Science[19] also exemplifies efforts at cooperative space activity. The IACG had its origins in a 1981 meeting in Padua, Italy of international space scientists. They wanted to coordinate planned scientific missions to explore Comet Halley. The scientists represented NASA, the European Space Agency (ESA), the Japanese Institute of Space and Astronautical Science (ISAS), and the then Soviet-led Eastern bloc group, Intercosmos. Their cooperative efforts culminated in the 1986 encounters with Comet Halley by five spacecraft, two from ISAS, two from Intercosmos, and one from ESA. Although the United States did not send a dedicated spacecraft, NASA involvement was essential for the Pathfinder Mission of the encounters, which provided crucial support from the US Deep Space Network to track the two Soviet spacecraft and subsequently provide information required to target the following ESA spacecraft as accurately as possible.

Although originally intended solely as a group for coordinating Halley's Comet efforts, the success of the IACG and its missions stimulated scientists to extend its duration. Consequently, the agencies involved unanimously decided to continue the IACG. Their new Terms of Reference reflected the group's basic philosophy: (1) there would be no or minimal technology transfer; (2) no funds would be exchanged; and (3) the group would only advise member agencies.

The IACG's objective is to maximize opportunities for multilateral scientific coordination among specific areas of mutual interest to all members. Its guidelines carefully and succinctly state the boundaries within which the IACG operates. Without limiting its scope, it would have been difficult or even impossible to obtain governmental support for the organization. A broader mandate would have moved its activities into more threatening political realms, and hence the group would not have survived.

The organization has, however, survived. Some have even suggested that the IACG be used as a precursor or model for a World Space Agency.[20] But the IACG has survived primarily because it is made up of scientists trying to work more effectively within the constraints of a political world. In Phase 2, wherein the IACG addresses the task of solar-terrestrial physics, it has learned some important lessons. The nature and scope of the Phase 2 project compounded the problems which would have been associated with the growth of any group.

Whereas the prospective Halley's Comet encounters seized the public's imagination and hence obtained political support (in other countries more than in the United States), solar terrestrial science is a basic science program of little interest to the general public or politicians. Phase 2 has suffered in terms of political commitment and subsequently resources. The lesson is that political interest is a two-edged sword: It can generate the financial support and managerial commitment to make a project successful, as was the case with IACG Phase 1. Or it can lead to bureaucratic scrutiny and/or neglect which reduces the chance of success.

It became glaringly evident at the 1990 IACG meeting in Tenerife that the IACG has expanded from a close knit, intimate group to one where formality replaces informality, and where commitment has not been offered commensurate with the complexity of the task. Growth and institutionalization is not always compatible with continued success. Indeed the IACG exemplifies the limitations of institutionalized cooperation in space. Space scientists view the IACG mechanism as desirable only for limited projects and far prefer to keep cooperation on a project-to-project basis. In non-scientific fields of possible space-related cooperation, political and economic issues rapidly become dominant and block institutionalized cooperation other than in anomalous cases such as the Space Station. Even in the case of the "internationalizing" the Space Station by including the Russians, "cooperation" has primarily been a means for the Clinton Administration to secure continued domestic funding for the Station and therefore fulfill a campaign promise, while reaping the political benefits of the Station's symbolism. But generally, space simply is not high enough on political agendas in the United States or other countries to merit the concerted effort that would be necessary to even begin thinking about broad-based institutionalized cooperation.

Space advocates looking for a "hook" on which to hang political support favor the environment: a low-politics issue of global interest, and one politically of particular interest to Vice-President Albert Gore. But linking space to non-proliferation concerns to raise the former issue's priority would not improve the situation. Rather, such linkage would complicate the problem of space's already low priority by elevating it to the realm of high-politics, decreasing the feasibility of institutionalized cooperation. Creation of a World Space Agency or World Space Organization which might monitor launch sites and hardware to slow or stop the transfer of missile technology, however desirable, has been evaluated as "premature" or simply not feasible. Kenneth Pedersen, for example, recently reviewed the prospects for a World Space Agency, first proposed by the Soviets in the mid-1980s. He concluded:

> I do not sense that the stars are aligned in favour of that event occurring in the near future, nor do I find this to be a distressing prospect. This judgment is based on my belief that the benefits of a WSA tend to be overestimated by its proponents and are likely, in any case, to be outweighed, at least in the near term, by several countervailing factors.[21]

This view was also reiterated in the report of a December 1992 workshop on international cooperation in space held in Hilo, Hawaii, sponsored by the American Institute of Aeronautics and Astronautics (AIAA):

> The notion of the creation of a "world space agency," a concept which has been raised by some in the recent past, was discussed but considered too formidable an undertaking at this time. Needed instead is a looser structure designed for political and programmatic flexibility which would allow the organization of one of more pilot projects to respond to global problems of special urgency.[22]

Finally, the issue arises of who would pay for such an organization. It has long been acknowledged that space transportation needs to move beyond ballistic rockets. A major space study has been done every year since 1986. Each stated that space transportation issue is high priority; yet nothing has been done because of the cost that would be involved. If garnering political support for badly needly hardware is difficult, then garnering political support to create another international space bureaucracy would be even more difficult, likely futile.

Conclusions

Four basic models of past cooperative efforts for the utilization or monitoring/control of technology have been offered for consideration: treaties, such as the NPT and the Outer Space Treaty; a formal international organization, such as the IAEA; an informal "regime" like the MTCR; and informal coordinating groups, such as the IACG and CEOS. Given that countries are highly unlikely to give up what is perceived to be in their national interest, a treaty or a formal organization reliant upon intrusive inspections to implement a cooperative space regime are unlikely to be well received by space-faring nations. The problems encountered by the IAEA (and the similarity of problems that would be encountered regarding missiles, particularly concerning inspections) and the lack of feasibility of a WSA all suggest that the formal organization model is unsuitable for non-proliferation efforts in the long range missile technology field. Tried and tested approaches such as the IACG and CEOS should be encouraged and incrementally strengthened, while also building on the capabilities of the MTCR.

My conclusion coincides with the AIAA workshop recommendation which stated that what is really needed to promote space cooperation is "a looser structure designed for political and programmatic flexibility." One way to begin moving forward would be to work within the established MTCR, and form special regional subsets of members to deal with parochial issues. This approach seems particularly suitable given that most proliferation potential results from local threats or regional conflicts. Although this approach would not address the motivations of all countries, it could be a start. Such regional efforts could periodically examine together common issues for the MTCR to deal with at a global level.

Cooperation in space activities will continue and increase, but not in the area of space transportation, and not under the rubric of formal organizations. The competitive market in space vehicles which already exists makes a cooperative effort along the lines of the INTELSAT model highly unlikely. Countries may seek to buy technology to upgrade existing systems. India purchased cryogenic rocket technology from Russia[23] only to have the contract cancelled later when Russian joined the MTCR; Russia balked at forgoing the $35 million contract, but was pressured by the United States into accepting MTCR guidelines in exchange for Russian inclusion in the Space Station.[24] Opportunities for technology purchase, therefore, may be more restricted than originally feared. Moreover, most countries are unable to afford "from scratch" development of an entire launch systems. Finally, the limits to consensual cooperation become obvious as soon as one asks: who would launch military-related satellites for countries like Libya, Iran, Iraq and South Africa? Economics may dictate one set of cooperation oriented efforts for space launches, but politics may well dictate another.

In short, if a country seeks missile technology, it will be able to obtain it. Space cooperation will continue, and expand in certain areas, but not in ways that differ much from that in the past two decades. Of course, windows of opportunity should not be ignored, but we should open them cautiously. National interests have not disappeared.

The entire situation could change if a next-generation launch vehicle were developed which enabled space commercialization to become a reality. If space transportation were to become privatized, much as the railroads and air travel did earlier, how would this innovation affect the diffusion of ballistic missile technology? Would reusable launch technology make expendable launch vehicles obsolete? Would all missiles become purely military? As cost has inhibited the development of such a new vehicles, perhaps funding needs of space transportation be addressed cooperatively, especially now, during the formative years of technology development.

Notes

1. W. McDougall, ...*the Heavens and the Earth,* New York: Basic Books, 1985, p. 123.

2. J. Johnson-Freese, "Nuclear Nonproliferation: U.S. and West Germany," *Journal of International Affairs,* Vol. 37/No. 2, Winter 1984, pp. 283-293.

3. W. Patterson, *Nuclear Power,* New York: Penguin Books, 1977, p. 242

4. W. Potter, "Nuclear Proliferation: U.S.-Soviet Cooperation," *The Washington Quarterly,* Winter 1985, p. 151.

5. W. McDougall, ...*the Heavens and the Earth,* p. 419.

6. N. Goldman, *American Space Law* (Ames, Iowa: Iowa State University Press, 1988), p. 91.

7. Ibid., p. 91.

8. NASA, Division of International Affairs, *26 Years of International Programs,* January 1, 1984, p. 2.

9. For a full discussion of this issue see: J. Johnson-Freese, *Over the Pacific: Japanese Space Policy Into the 21st Century* (Iowa: Kendall-Hunt Publishing, 1993), Chapter 8.

10. These criteria are discussed in: J. Johnson-Freese, *Changing Patterns of International Cooperation in Space* (Melbourne, FL: Kreiger Publishing, Orbit Books, 1990), Chapter 1.

11. A. Frye, "Zero Ballistic Missiles," *Foreign Policy,* Number 88, Fall 1992, p. 3.

12. Ibid., p. 16.

13. Ibid., p. 14.

14. European Space Agency, *Reaching for the Skies* (Paris, Cedex 15, France, 1988) BR-42, p. 14.

15. "Mitsubishi Leads Privatization Effort As First H-2 Boosters Are Fabricated," *Aviation Week & Space Technology*, August 13, 1990, p. 64.

16 M. Dornheim, "Costly High-Tech Projects Could Hurt Japan's Commercial Space Competitiveness," *Aviation Week & Space Technology*, March 23, 1992, p. 73.

17 A. Karp, "The Commercialization of Space Technology and the Spread of Ballistic Missiles," in *International Space Policy*, eds. D. Papp and J. McIntyre, New York: Quorum Books, 1985, pp. 179-194.

18. American Institute of Aeronautics and Astronautics, *International Space Cooperation: Learning from the Past, Planning for the Future*, 1993.

19. For a more detailed description of the IACG and its activities see: J. Johnson-Freese, "A Model for multinational space cooperation: The Inter-Agency Consultative Group," *Space Policy*, November 1989, pp. 288-300; and "The Evolution of the Inter-Agency Consultative Group: From Halley's Comet to Solar Terrestrial Science," *Space Policy*, August 1992, pp. 245-255.

20. K. Pedersen, "The Global Context: Changes and Challenges," *Economics and Technology in U.S. Space Policy,* ed. Molly Macauley, Proceedings of a Symposium held in Washington, D.C., June 24-25, 1986, Resources for the Future and the National Academy of Engineering, p. 187.

21. K. Pedersen, "Is It Time to Create a World Space Agency?" *Space Policy*, May 1993, p. 91.

22. *International Space Cooperation,* Report of an AIAA Workshop, AIAA, Washington, DC, March 1993, p. 50.

23. V. Raghuvanshi, "Yeltsin: Cryogenic Rocket Deal Is Irrevocable," *Space News*, February 1-7, 1993, p. 6.

24. D. Sneider, "Russians Up in Arms About Cancellation of Rocket Deal, *Christian Science Monitor*, July 27, 1993, p. 7.

7

Exchanging Environmental Resource Management for Peaceful Space Practices: Blue Sky or "Blue Sky"?

Molly Macauley

Concern about defense management (particularly nuclear proliferation) has long figured prominently in international relations involving countries in the Middle East and South and East Asia, and more recently, with nations in Central and East Europe. Meanwhile, concern about the environment has also increased in international prominence, most recently culminating in the United Nations' Conference on Environment and Development (the "Earth Summit") held in June 1992. In light of these twin concerns, some experts have suggested that useful linkages might be made between environmental and nuclear proliferation issues in structuring the terms of international lending and other assistance policies.[1] Indeed, for example, as Central and Eastern Europe attempt a transition to more market-oriented economies, prominent concerns range from the ownership and management of defense weapons and their possible conversion to civilian applications (production facilities as well as some components, such as launch vehicles) to environmental problems posing substantial health risk. Thus, at first glance, it seems as if linkages between these concerns might be practicable.

In this paper, I generally argue against the feasibility of policies linking the environment and defense, although I do consider one activity, remote sensing, as a possibly promising link between space launch (and other space-based programs) and environmental resource management. I proceed as follows. First, I discuss the difficulty of linking environmental concerns and defense management, essentially because environmental issues among countries where proliferation

119

concerns loom largest are overshadowed by substantial issues of poverty and economic growth. In short, environmental problems receive low priority.

Next, I consider remote sensing as a possible exception to this logic in that it may provide an attractive link between space and environmental activities. The problem of the low priority accorded environmental concerns remains, however. And another problem arises--that of national autonomy in access to remote sensing technology.

In part three, I first discuss some of the theoretical concepts that have been developed (primarily in the economics, political science, and game theory literature) for linked compensation, and then consider these concepts in the context of proliferation and environment. The presumption underlying this discussion is that when and if environmental concerns become elevated among national priorities (that is, when the concerns expressed in section II are alleviated), then a fuller context for considering linked compensation will be useful for forming policy.

The Difficulty of Linking Environmental
and Defense Management: Mismatched Priorities

The principal reason why policy prescriptions linking environmental management with defense activities may not work is that environmental protection is generally accorded secondary (at best) importance or, more typically, even lesser importance in non-western economies.   Urgent environmental concerns tend to be those directly linked to health, such as water quality and sanitation. Yet even these concerns are generally inseparable from broader economic issues related to the alleviation of poverty and the fostering of economic development.[2]   This status contrasts sharply with that accorded the environment in the United States, where environmental quality has been treated as an autonomous objective since the 1970s (when legislation such as the Clean Air and Clean Water Acts were passed). In part, this result is because demand for environmental quality tends to be highly income elastic--that is, demand grows with growth in income, much like demand for leisure goods.

Missile proliferation concerns are concentrated among nations in the Middle East, South and East Asia (see Nolan and Wheelon, 1990), and members of the Commonwealth of Independent States (Ukraine, Belarus, Kazakhastan, and Russia). Environmental concerns generally figure much less prominently than other domestic and international concerns in these areas. As some experts have emphasized:

The former Soviet Union and the countries of eastern Europe have
shown less enthusiasm for CO_2 emissions reductions than the OECD

countries. They are preoccupied with reforming their political and economic systems and addressing pressing local environmental problems...Developing countries are perhaps the least eager to support an international agreement to curb CO_2 emissions because they fear it will have a negative impact on their economic development efforts...In any case, the developing countries believe that their contribution to global warming is being overstated and that the developed countries created the problem and should assume responsibility for mitigating it.[3]

In fact, some environmental prescriptions conflict directly with the livelihood of potential proliferators. In debate on controlling greenhouse gas emissions, for example, Middle East nations (largely oil exporters) tend to argue against carbon controls unless special treatment, such as compensation, is accorded their countries. In East Asia, Taiwan and South Korea rely extensively on nuclear-generated electricity, as do members of the Commonwealth of Independent States (CIS), perpetuating the concern between the potential production of weapons-grade fuels and fuels used for power production. Moreover, stepped-up nuclear power production has figured prominently in some plans to meet the needs of increasing industrialization, thereby complicating this facet of an environment/proliferation link. In eastern Europe, the response to gross inefficiencies in energy use and rising energy prices, coupled in some cases with coal mining as large sources of employment, has not been to improve energy efficiency or adopt clean coal technologies. Rather, the response has been to raise wages and increase the money supply to avoid severe short-run macroeconomic dislocation (at the cost, however, of severe inflation and set-backs in long-run economic adjustment).

This discussion is not intended to conclude that potential proliferators are unanimously relegating environmental concern to the lowest of priorities. There are growing green movements in India; for example, Pakistan attended the Earth Summit with a list of environmental concerns. South Korea has banned leaded gasoline and instituted new environmentally-related disclosure rules on chemical imports.[4] In addition, environmental issues in these countries (and others), such as population growth, fossil fuel use, and priority areas for conservation of biological diversity have implications for international resource management and thus attract the attention of western economies. What is less clear, however, is the willingness on the part of the western economies to pay for improved environmental management in these countries (more on this in the following sections).

Moreover, if history is any guide, environmental concerns have usually diminished in importance when, as is often the case, they are entangled in political conflict. For example, the entire Eastern bloc boycotted the first global conference on the environment, the United Nations' Conference on the Human Environment (held in Stockholm in June 1972), because of extant political conflicts over the postwar division of Germany.[5] Presently, the hesitation of some CIS members (most notably, Ukraine and Kazakhastan) to agree to non-proliferation pacts has held up economic assistance (environmental and otherwise) that might be expected from the United States. Moreover, popular sentiment pushed Ukraine to keep the weapons and to hold out for financial compensation (estimated to be $1 billion) for dismantling the warheads. The symbolic (as well as actual) value of nuclear armament was also viewed as integral to Ukraine's independence, ability, and readiness to defend itself.[6] Similar sentiment has been growing in South Korea, wary of North Korea's stance on nuclear weapons.[7]

Environmental concerns may become more important with time due to their impact on economic growth. Typically, environmental expenditures are seen as drains on the economy rather than additions to its productivity. If productivity increases that are associated with environmental improvements are potentially large, however, and if decision makers and voters perceive potential gain from these improvements, then a joint treatment of environmental and defense-related issues may be able to transcend political concerns. Yet very little research has been conducted to quantify the total productivity-related impacts of environmental degradation on the economies of developing countries or those of members of the CIS (in terms of lost productivity and other measures related to economic growth).[8]

A Possibly Feasible Link

One possible indirect linkage between the environment and space is that which might naturally arise from the civil and commercial potential of military assets. Some experts have suggested (while noting that the proposal is ambitious) a "club" to provide launch facilities or a "common carrier" for peaceful space missions, perhaps providing satellite reconnaissance data to all members.[9] This action in turn could be linked to other observers' suggestions for international environmental institutions[10]--for example, the sharing of satellite data for environmental monitoring and resource mapping. That there is some interest in these approaches is suggested, for example, by Ukraine's national space agenda, which includes the launch of remote sensing spacecraft for ocean study and environmental monitoring and resource mapping.[11] Institutionally, there is precedent for such activity. Provisions of arms reduction treaties include as a means of destruction the use of missiles for peaceful purposes, such as launching civil research payloads.   Russia reportedly launched an experimental communications satellite on a converted SS-25 ballistic missile in March, 1992.

Ukraine has proposed to the United Nations' Committee on the Peaceful Uses of Space to use ICBMs for commercial launch.[12] Yet, the U.S. Department of Defense has recommended against the use of excess U.S. ballistic missiles as orbital space launchers (suborbital research missions appear to be permitted in the draft policy) except for government purposes, to be decided on a case-by-case basis. In particular, the Department has recommended against commercial use of these assets (apparently responding to pressure from the commercial launch industry that government assets not be released to the market, thereby depressing launch prices).[13]

There is also a long track record of international cooperation in space, including sharing environmental and weather data, as might be envisioned in a policy linking peaceful uses of launchers and the development of space-based civil remote sensing satellites.[14] At the same time, however, there is prominent competition in space, partly to acquire prestige and to demonstrate technical prowess through autonomous space capability, including remote sensing. Remote sensing not only demonstrates technical acumen, but also conveys control over surveillance activities whether for defense or environmental observation, monitoring, and measurement. Thus, unlike other types of space activities, remote sensing programs are essentially information gathering activities, and information brings with it a sense (if not the realization) of control. For this reason, nations may not be willing to share in global remote sensing consortia, or may be willing to do so while at the same time sponsoring their own autonomous programs. These reasons may explain why as many as 30 nations are expected to have their own remote sensing spacecraft systems by the end of the decade, despite the tradition of sharing meteorological and other environmental data.[15]

A related difficulty in linking environment and space capability by way of remote sensing is that the highest priority environmental problems in potential proliferating states do not seem to coincide with the types of environmental indicators observable from space. For example, water quality and sanitation are high priority, as are introducing unleaded gasoline and improving energy efficiency. In central and eastern Europe, airborne lead and lead soil contamination, sulfur dioxide emissions, and nitrates and other contaminants in water are among the issues which are highest on the environmental agenda. Although land management (of wetlands, coastal ecosystems, mountain habitats), and in general, the preservation of biodiversity are also listed, they are of somewhat lower priority.[16]

Environmental indicators most directly monitorable from space observations are not necessarily related to these activities; rather, space-gathered diagnostics are typically related to upper atmospheric diagnostics (such as ozone monitoring) and the somewhat lower-priority land use issues. The best measures of sulfur dioxide emissions, for example, are *in situ* measures of the sulfur content of feedstock coals and unburned ash, and emissions monitoring devices installed in the smokestacks.[17]

Finally, some of the difficulties associated with defense conversion or implementing dual use technologies (for civilian and defense purposes, and in reconfiguring launchers or broader conversion activities) should not be underestimated. Some defense conversion is recognized as economically necessary in CIS countries, for example. Yet the political will to complement this policy seems lacking. A former U.S. aerospace official has commented that there are no examples of successful conversion in the narrow sense of changing an existing plant and production team working for the military market, and redirecting them to produce for the civilian market.[18] Moreover, successful military conversion depends upon successful economic and institutional reform--macroeconomic stabilization, currency convertibility, the transition to private property, and the challenge of adopting and implementing practices of a market economy.[19] In this regard, both military conversion and environmental concerns are situated in the broader context of political and economic transition. Thus, the feasibility of a "natural" link between reducing proliferation potential and remote sensing projects, in particular, seems to be somewhat tenuous.

Some Conceptual Arguments for Linked Compensation

Despite the discouraging tenor of the above arguments against linking improved environmental management with incentives for non-proliferation, there are several conceptual bases for linking compensation--that is, compensation that is directly targeted at an activity. These bases may become useful for policy debate as environmental concerns become more prominent (say, if the problems worsen, or if the links between environment and welfare can be demonstrated convincingly). The theoretical literature has discussed these bases in the context of international environmental issues. Is it instructive in the case of joining the environment and non-proliferation in approaches to international policy? In this section, I review these conceptual bases and discuss their application to the environment/proliferation issue.

D. Burtraw and M.A. Toman offer a good overview of rationales for linked compensation in the case of compensation from industrialized to developing countries to address climate change.[20] General tenets of welfare economics suggest that it is best to give money rather than goods and services and let recipients spend the funds as they choose. This approach maximizes the opportunity for recipients to improve their lot as they see fit, and avoids the information gap that may make compensation in forms other than cash of less use. For example, cash transfers rather than food stamps (or a negative income tax) have been advocated on these grounds in domestic economic policy. In the case of environmental policy, in its discussion of mechanisms for implementing international environmental agreements, the World Bank has urged that "transfers should take the form of lump-sum payments rather than finance for specific investments."[21]

In several cases, however, linked compensation of in-kind rather than cash resources may better serve the interests of both donors and recipients. Following Burtraw and Toman, these include:

1. Linked compensation may be less expensive for the donor than financial aid. An example is the transfer of technology from the donor country, in which case industry in the donor country may obtain economic side benefits from the extra business, lowering the opportunity cost of the assistance compared with direct financial aid. The transferred technology may also permit a sharper targeting of the assistance to problems of mutual concern, such as transboundary pollution from the recipient country to the donor country(ies).
2. In-kind compensation may not be as easily transferred as financial aid, thus reducing the potential for adverse selection (when recipients misrepresent their qualifications for the aid).
3. In-kind compensation may also reduce any tendency for moral hazard (when recipients act to attract more aid, as might be the case if recipients delay the installation of pollution control devices in order to increase aid).
4. In-kind compensation might be preferred by recipients if it attracts less political reaction or is less susceptible to misappropriation by political or military elites in the recipient countries.
5. In-kind compensation may encourage recipients' participation if it appeals to a notion of fairness (for example, if developing countries were to receive technical assistance for pollution problems felt to be caused by industrialized countries).
6. It can be difficult to calculate the appropriate level of direct monetary compensation, and it can be construed as a "buy off" of concerns not taken seriously.

How do these general arguments for linked compensation fare in considering the exchange of environmental assistance in return for non-proliferation? With respect to the first argument, whereby donors may obtain side benefits, such as expanded markets when aid is given in-kind, the argument could be made in relation to an exchange of environmental technology. For example, $1.5 to $3 billion worth of pollution control equipment could be given to Ukraine as compensation for dismantling its missiles. There are several problems with such a proposal, however. One problem is that pollution remediation is not among the highest priorities for Ukraine; rather, obtaining hard currency is at the top of the list. In addition, it would be challenging politically to balance the other arguments for retaining nuclear capability (defense, symbolic value) against the benefits of environmental protection. Finally, the actual amount of side benefit potentially accruing to donor industries from an exchange of pollution devices may not be associated with large net profits, as much of the recommendations for environmental control center on "low technology" solutions, which may not have higher profit margins.

The second and third arguments, relating to adverse selection and moral hazard, are particularly weak in the case of exchanging environmental control for non-proliferation. The exchange of technology for dismantling weapons *would* support these arguments, as the technology would be directly related to the activity in question. The exchange of environmental technology would not support them, however, as it would not be linked to the weapons' dismantlement activity. Thus control over the activity would not be strong. If the environmental technology related specifically to the use of space assets--such as remote sensing for resource management and mapping, or use of launch vehicles for civil space science or peaceful commercial purposes related to the environment[22]--then these arguments could be affirmed. Again, however, the difficulty is elevating environmental concerns to some status more closely aligned with that given to defense conversion in general, and proliferation capability in particular.

The fourth argument, suggesting that in-kind compensation might be more politically attractive and less susceptible to misappropriation, also would require heightened popular and political environmental concern in proliferant states before it could be said to support an environment/missile tradeoff. This result is especially the case to the extent that some types of environmental remediation may imply significant job loss (such as reducing coal use, in which case large mining regions, such as the Donets'k basin in Ukraine, could suffer large unemployment). In this case, the political stakes are quite high rather than low. As in the preceding arguments, direct exchange of technology for non-proliferation might be preferred if misappropriation is a concern.

The fifth argument bifurcates the spectrum of regions where proliferation is a concern. That is, the argument seems to apply differentially to developing countries in the South compared with countries in central and east Europe. The South (generally) tends to cast its pollution concerns from the perspective of "victims" or recipients of pollution damages. Central and east Europe, by contrast, have tended to contribute to as well as be the victims of pollution. Accordingly, the "fairness" notion may apply more in the former case than in the latter. The fairness notion also arises with who should pay for dismantling weapons; for example, requests for financial compensation to dismantle are generally supported by appeals to equity for the common good.

The sixth argument, reflecting the difficulty of calculating the level of direct monetary compensation, is less worrisome if the calculation involves the cost of pollution control devices and the cost of dismantling weapons--both activities can presumably be measured using conventional cost accounting applied to the technologies required in both cases. On this basis, the relative costs of arms control verification investment seem much smaller than even the more conservative estimates of the cost of environmental remediation among developing and CIS countries.[23] More fundamentally, however, the challenge lies on the benefit side, not the cost side. Calculating benefits of improved environmental quality is difficult, as is, certainly, calculating the benefits of controlling nuclear proliferation. As both benefits and costs will undoubtedly color the bargaining perspectives of negotiating parties (implicitly if not explicitly), these difficulties should not be underestimated.

The first, fourth, and fifth arguments are the general tacks now being taken by lending agencies and western nations in the case of environmental assistance. Observers at the recent ministerial level conference on environmental issues in central and east Europe noted, however, that "Environment ministers from Russia, Slovakia and other nations said that environmental issues, for all their gravity, had moved far down on domestic agendas because of the immediate problems of jobs, food and financing."[24] Accordingly, it seems clear that raising the priority given environmental concerns depends on the extent to which they can be linked to improving economic productivity--and little is known about the size of this link. Burtraw and Toman note that compensation to developing nations could be targeted toward general economic development or population planning, as these are inextricably linked to quality of the environment. Similarly, defense conversion is linked to economic conditions (employment, trade) as well as to political concerns (national sovereignty, autonomy, demonstration of technological prowess). What improved policy design may need, then, is better information about the economic impacts (the benefit side) of environmental and defense policy, to better understand relative priorities and potentially beneficial linkages between them.

Conclusions

Because the priority given to environmental concerns seems so low among countries where nuclear proliferation is a concern, it is probably unlikely that an exchange of environmental assistance for non-proliferation activities is likely to be attractive. Moreover, the costs of environmental remediation--even the lower estimates offered by the World Bank--seem quite large compared with the direct costs of some arms control verification investments, which taken alone argues for direct spending on verification actions rather than linking compensation. Of course, the benefit side of the equation in comparing these alternatives also needs to be taken into account--the efficacy of verification investments, and whether the relation of environmental degradation to economic growth can be shown to be large and significant (that is, if environmental management can be shown to be complementary rather than deleterious to economic prosperity).   If the latter dominates the former (that is, if net environmental gain is larger than the net gain in the effectiveness of verification actions), then a more convincing case can be made to elevate the environment to greater prominence in international debate. In turn, such evaluation could promote the attraction of remote sensing programs as an alternative use for space assets. Otherwise, the more direct approaches to non-proliferation--improved verification, challenge inspections, the sharing of intelligence[25]--may be the more fruitful strategy.

Notes

1. For example, see E. Shervadnaze, "Let's Trade Arms for Ecology," *World Press Review,* November 1991, v. 38, no. 1, p. 60; M.K. Tolba, "Environment for Peace," *Futures,* September 30, 1990, pp. 764-65; and J.C. Clad and R.D. Stone, "Green Foreign Aid Would Sell Better," *The International Herald Tribune,* May 5, 1993.

2. For example, see discussion in P.M. Haas, M.A. Levy and E.A. Parson, "Appraising the Earth Summit," *Environment,* 1992; World Bank, *World Development Report: Development and the Environment* (New York: Oxford University Press 1992).

3. P.M. Morrisette and   N.J. Rosenberg, "Climate Variability and Development," in J. Darmstadter, ed., *Global Development and the Environment: Perspectives on Sustainability,* Washington, DC: Resources for the Future, 1992, p. 77

4. See F. Hoke, "India Rejects 'Green' Courts," *Environment,* November 1990, v.32, no. 9, p.21; R. Jenkins, "Green Democracy in India," *Progressive,* May 1990, v. 54, no. 5, pp. 15-16; F. Bokhari, "Pakistan Heads to Brazil with its Own Eco-Worries," *The Christian Science Monitor,* June 1, 1992, p. 6; I. Breskin, "Korea Toxics Law Imperils Exports of U.S. Chemicals," *Journal of Commerce,* v. 393, July 27, 1992, p 1A; and *The Oil Daily,* December 23, 1992, v. 92, p. 5.

5. Haas, Levy, and Parson, 1992.

6. See discussion in "Ukraine: You'd Be Nervous Living Next to a Bear," *The Economist,* May 15, 1993, pp. 21-23; "How to Handle Ukraine? Sweetly!" *The Baltimore Sun,* May 15 1993; N.A. Feduschak, "U.S. Stance on Ukraine

Pushes It Closer to Declaring Itself a Nuclear Nation," *The Wall Street Journal,* May 3, 1993; "Heels Dug In," *The Economist,* January 9, 1993 pp. 43, 46; and S. Coll, "Nuclear Goods Traded in Post-Soviet Bazaar," *The Washington Post,* May 15, 1993.

7. See D.E. Sanger, "Wary of North, Seoul Debates Atomic Bomb," *The New York Times,* March 19, 1993, and K.D. Jung, "The Once and Future Korea," *Foreign Policy,* Spring 1992, pp. 40-55.

8. These *are* studies of *sectoral* economic impacts, but few have implemented a full social net benefit accounting focuses on overall productivity. See Krupnick, in Darmstadter, 1992, for discussion of the usefulness and drawbacks of using benefit-cost analysis to set priorities for environmental problems in developing countries. Krupnick sets forth a general framework, and then discusses key issues in its application to developing countries: the basic tenet of individual sovereignty underlying benefit-cost analysis; the influence of poverty on valuation; and the use of money as a common numeraire but not the sole criterion for policy evaluation.

9. For example, see J. E. Nolan and A.D. Wheelon, "Third World Ballistic Missiles," *Scientific American,* v. 263, no.2, August 1990, pp. 34-40.

10. For example, see Shevardnadze, 1991, and Tolba, 1990.

11. See Joint Publications Research Service, "Director of Ukrainian Space Agency Interviewed," *JPRS* USP 92 006, November 17, 1992, p. 35 (from Vecherniy Kiyev, August 6, 1992, p. 2).

12. Ibid.

13. See B. Iannotta, "DoD Recommends Tight Restrictions for Use of Missiles as Launchers," *Space News,* May 10-16, 1993, v. 4, no. 19, p. 1.

14. For discussion of success in cooperation, see National Aeronautics and Space Administration, *Twenty-Six Years of NASA International Programs,* (Washington, DC: U.S. Government Printing Office, 1984); D.T. Lauer, "Role of International Cooperation in Civilian Satellite Remote Sensing," Sioux Falls, SD, U.S. Geological Survey, EROS Data Center, 1990 (mimeo); and National Research Council, *Space Science in the Twenty-First Century: Mission to Planet Earth* (Washington, DC: National Academy Press, 1988).

15. See discussion of the tension between national sovereignty and remote sensing data in, for example, J.H. Ausubel and D.G. Victor, "Verification of International Environmental Agreements," 1991 (mimeo); M. Macauley, "Collective Goods and National Sovereignty: Conflicting Values in Global Information Acquisition," *Space Monitoring of Global Change,* (conference proceedings: University of California, San Diego, Institute on Global Conflict

and Cooperation and the California Space Institute, 1993); and *Space Business News*.

16. See World Bank, 1992, and "Executive Summary," *Environmental Action Programme for Central and Eastern Europe,* April 5, 1993 (mimeo).

17. Among the most comprehensive analyses of the correspondence between space monitoring capability and environmental diagnostics, see Ausubel and Victor, 1991, and B.D. Berkowitz, "The Use of Intelligence Resources for Environmental Monitoring," Washington, DC, Council on Foreign Relations, 1993 (mimeo). Berkowitz offers a table mapping the type of intelligence (for example, electro-optical imagery, synthetic aperture radar imagery, signature intelligence) into environmental data requirements. See also J. Wettestad, "Verification of International Greenhouse Agreements: A Mismatch between Technical and Political Feasibility?" *International Challenges,* 1991, v. 11, no. 1, pp. 41-47, for discussion of use of satellites for monitoring forest resources and deducing greenhouse gas emissions in part from the satellite forestry data; and J. Lanchbery, O. Greene, and J. Salt, "Verification and the Framework Convention on Climate Change," 1992 (mimeo); and W. Fischer, J.C. di Primio, and G. Stein, "A Convention on Greenhouse Gases: Towards the Design of a Verification System," October 1990, for discussion of satellite monitoring of forest and agriculture resource management and associated environmental impacts.

18. B. Hogan, "Conversion '91: Military Conversion Conference: Summary," Cambridge, Mass.: Harvard University, 1991, p. 3 (mimeo).

19. Hogan, 1991.

20. D. Burtraw and M.A. Toman, ""Equity and International Agreements for CO_2 Containment," *Journal of Energy Engineering,* August 1992, v. 118, no. 2, pp. 122-135.

21. World Bank, 1992, p. 156.

22. As discussed in Nolan and Wheelon, for example.

23. For example, the World Bank estimates that $75 billion annually could redress the most important environmental problems among developing countries; the estimated full cost of implementing the Earth summit's agenda for achieving sustainable development is about $600 billion annually. By contrast, the estimated cost of arms control verification measures by the U.S. range from about $3 billion in start up costs plus $600 million annually in recurring monitoring and inspection costs, to $5 to $15 billion a year if a steady-state constellation for 15 to 32 imaging and surveillance satellites is established (J.D. Morrocco, "Initial Compliance Costs for major Arms Treaties Could Exceed $3

Billio, Plus $660 Million Yearly," *Aviation Week and Space Technology,* November 19, 1990, pp. 74-75). The estimated cost of dismantling Ukraine's ICBMs, as noted earlier, is $1.5 to $2 billion. Based on these figures, spending on verification and dismantling rather than on environmental problems seems the less expensive option if the objective is nonproliferation. But in neither case is there evidence of much willingness to pay--in the case of the environment, western economies have contributed only a few billion through lending institutions or direct grants; in the case of dismantling, the United States, for example, has offered a "down payment" of $175 million toward Ukraine's dismantling activities.

24. M. Simons, "West Adopts Limited Plan to Clean Up East Europe," *The International Herald Tribune,* May 5, 1993.

25. For example, see discussion in J.S. Nye, Jr., "New Approaches to Nuclear Proliferation Policy," *Science,* May 29, 1992, v. 256, pp. 1293-1297; and Nolan and Wheelon, 1990. These approaches could include the suggestion in Nolan and Wheelon for a "space club" where space activities might be more transparent, and where the price of admission is transparency rather than "giving up" technology.

8

Arms Control and Verification: Future Trends

M. Lucy Stojak

The dual use (civilian/military) of almost all space capabilities raises concerns over the proliferation of, particularly, launch capability. Indeed, ballistic missile and space-launch vehicle technologies are virtually identical.

Currently, seven states possess an indigenous space-launch capability (the United States, CIS/Russia, China, France, Japan, India and Israel), while a number of countries are developing independent orbital launch capability (Brazil, Pakistan, South Korea). The lucrative business of arms trading is encouraging certain states to supply ballistic missiles (BM) to other countries. The collapse of the Soviet Union brings an added danger to ballistic missile proliferation: the possibility that advanced weapons, technology and knowledge get sold to unstable regions for economic reasons.

Yet, international cooperation of various kinds has become a central feature of the world's space programs. Cooperation with launching states gave countries that did not possess their own access to space the opportunity to participate in space science and applications efforts. The international legal framework applicable to outer space reflects the international community's desire to ensure that the exploration and use of outer space be carried out for the benefit and in the interests of all countries.

How then does one create a regime for outer space which makes allowance for the dual-nature of space technologies while advancing the use of applications which promote stability? The difficulties in distinguishing between ballistic missile and spacelaunch vehicle testing has renewed calls for on-site inspection of payloads. This paper reviews past proposals for on-site inspection, as well as institutional arrangements discussed in the United Nations (UN) and the Conference on Disarmament (CD). It argues that the prospects for a multilateral agreement on dual-use technologies are poor. In light of present constraints, measures aimed at building confidence in outer space activities and the use of its technologies are more pragmatic and stand a better chance of being accepted.

The Historical Arms Control Context

Even prior to the birth of the space age with the successful launch of Sputnik I, the question of ensuring that objects sent through outer space would be used for exclusively peaceful purposes, prompted the US delegate to the UN to state that the first step toward the objective of ensuring that future developments in outer space be devoted to peaceful and scientific purposes is "to bring the testing of such objects under international inspection participation."[1]

In a recent study carried out by UNIDIR it was stated that the most reliable and effective way of distinguishing between missile testing and space launches "would be on-site, pre-launch inspection of all objects, both civilian as well as military destined for outer space."[2] More recently, proponents of a Zero Ballistic Missile (ZBM) regime have also advocated pre-launch on-site inspection of payloads for distinguishing between such testing.[3]

On-site inspection has been a bone of contention in arms control and disarmament negotiations almost from the beginning of the space age.

The on-site launch inspection concept emerged in 1961 and found expression in the provisions regarding an International Disarmament Organization (IDO), itself part of an overall initiative regarding general and complete disarmament. Although the IDO concept never came to fruition, a number of its provisions addressing inspection of objects to be launched in outer space are worth recalling.

Both the United States and USSR submitted draft treaties on general and complete disarmament to the Eighteen Nation Disarmament Committee (ENDC).[4] The Soviet draft envisaged the creation of an IDO with its own internationally recruited staff.[5] The personnel of the IDO was "to enjoy in the territory of each state party to the Treaty such privileges and immunities as are necessary for the exercise of independent and unrestricted control over the implementation of the...Treaty."[6]

The US draft adds that the IDO and its inspectors would control all rockets and other space devices to ensure their use for peaceful purposes, through "inspection teams at the sites for peaceful rocket launchings who shall be present at the launchings and shall thoroughly examine every rocket or satellite before their launchings."[7] The Soviet proposal contained similar wording.[8]

This undertaking was to be carried out during Stage I of the General and Complete Disarmament process which involved the destruction of "all rockets capable of delivering nuclear weapons" (and their launch sites) "except for an agreed number to be retained by the USSR and the United States until completion of Stage II." Verification of this would be carried out by the IDO through launch site inspections. In addition:

> The production and testing of appropriate rockets for the peaceful exploration of space shall be allowed, provided that the plants producing such rockets, as well as the rockets themselves, will be subject to supervision by the Inspectors of the International Disarmament Organization.[9]

It is clear from the foregoing that both the United States and USSR considered on-site inspection of spacecraft as a quintessential component of a general and complete disarmament plan. The main focus however was on the control of launch vehicles and launches of nuclear weapons on ICBMs. No mention is made of on-site pre-launch payload inspection. At that time, both countries were actively developing their satellite reconnaissance capabilities, and did not want highly sensitive technology to be inspected. Finally, the details of how these inspections would be carried out were never greatly elaborated. Even during their preparation it was recognized that the drafts were more utopian than pragmatic.[10]

With the rejection of general and complete disarmament as a viable goal, on-site inspection disappeared from the diplomatic world stage for a generation. The 1960s and 1970s witnessed technological innovation, notably in the field of satellite reconnaissance and seismic sensors. These advances led to the development of "national technical means of verification" (NTMs). The development of NTMs enabled both super powers to successfully negotiate several key bilateral arms limitation agreements. Because this form of verification is "peripheral"[11] and involves no derogation from or compromise of sovereignty, the advent of NTMs helped the United States and USSR "to disentangle themselves from the rigid unyielding debates of the 1950s over on-site inspection."[12]

Recent Developments in Arms Control

With the East-West rapprochement, a climate of political acceptability for on-site inspection in the arms control arena has recently emerged. In the several strands of negotiations, conventional, chemical and nuclear, on-site inspection has been incorporated into the texts of international agreements notably, the *INF*

Treaty,[13] the *CFE Treaty,*[14] the *Vienna Document of 1992*[15] and the *Convention on Chemical Weapons.*[16] The recently signed *Open Skies Treaty*[17] provides for aerial observation. All of these treaties clearly represent derogation from the principle of State sovereignty.

Aimed at eliminating an entire weapon system, the 1987 INF Treaty contains such wide-ranging verification measures as short-notice, on-site inspections, continuous portal monitoring of missile production facilities[18], and detailed exchange of information. Notably, except for continuous portal monitoring, missile production facilities are not subject to any on-site inspection on the insistence of the United States, which invoked reasons of "national security."[19] These measures are supplemented by the use of NTMs.[20]

The INF Treaty provides for data exchange and notification through the Nuclear Risk Reduction Centres (NRCC).[21] Worthy of note is that "each party may, at its own discretion as a display of good will and with a view to building confidence, transmit through the NRCC communications other than those provided for under article 1 of this Protocol."[22]

The 1990 CFE Treaty will bring about the first large scale reductions in conventional forces in Europe (CFE) covering the area from the Atlantic to the west of the Urals (ATTU). Thousands of pieces of equipment such as tanks, armored combat vehicles, artillery, combat aircraft and helicopters will be withdrawn and destroyed. States parties may also use NTMs or multinational technical means (MTM) of verification.[23] This language in the CFE Treaty opens the possibility for closer European cooperation in monitoring from space. As a first step, closer cooperation is foreseen in the framework of the Western European Union (WEU) which is to start operating a satellite data analysis center using images from commercial satellites like SPOT and LANDSAT, as well as images from the military satellite HELIOS, a joint French-Spanish-Italian satellite, to be launched in 1993/94. Eventually, data obtained by Soyuzkarta could also be included.[24]

The CFE Treaty provides for on-site inspections of declared and undeclared sites, to reduction sites[25] and to certification sites.[26] Unless the quotas are exceeded, there is no right of refusal for on-site inspections of declared sites. There is, however, a right of refusal and a right of delay for challenge inspections. Finally, pursuant to Article XIV (6), "each state party shall have the right to conduct, and each state party with territory within the area of application shall have the right to accept, an agreed number of aerial inspections within the area of application." No agreement could be reached on the details of overflights, but the state parties are committed to negotiating them in the follow-on talks. (Art. XVIII).

Information to be exchanged under the CFE Treaty is contained in the Protocol on Notification and Exchange of Information. Each state party shall provide to all other states parties the information specified in the Protocol in accordance with the procedures set forth in the Annex on Format.

The Annex states that the information in each data listing shall be provided in mechanically or electronically printed form.

Finally, states parties shall use diplomatic channels designated by them, including a communications network to be established by a separate arrangement.

The 1992 Vienna Document applies to all the participating states of the Conference on Security and Cooperation in Europe (CSCE). It is part of the CSCE process.

In 1983, the Concluding Document of the Madrid Meeting of the CSCE urged states to undertake negotiations on the adoption of new confidence and security-building measures (CSBM) that would be militarily meaningful and politically binding and provide for adequate forms of verification. Such measures were intended to increase transparency in the military sphere. The CSBMs thus agreed upon were consolidated in the 1986 Stockholm Document. They were subsequently supplemented and expanded upon by further CSBMs together with which they formed the 1990 Vienna Document. The latter has now been superseded by the 1992 Vienna Document, which again incorporates new CSBMs.[27]

The CSCE states also agreed in 1990 to establish a CSBM data bank, to be administered by the Conflict Prevention Center in Vienna.

Pursuant to the 1992 Vienna Document, participating states will exchange information on their military forces, data relating to major weapon and equipment systems and their deployment, as well as information on military budgets.[28]

Information provided under the provisions on information on Military Forces and on Plans for the Deployment of Major Weapon and Equipment Systems will be subject to evaluation. Each participating State will provide the opportunity to visit active formations to allow the other participating states to evaluate the information provided.[29]

The Treaty on Open Skies establishes a regime to improve openness and transparency, to facilitate the monitoring of compliance with existing or future arms control agreements and to strengthen the capacity for conflict prevention and crisis management in the framework of the CSCE and other relevant international institutions.[30]

Under this Treaty, signatory states shall have the right to conduct flights over the territory of other states parties and shall be obliged to accept flights over their own territories.[31] Such observation flights shall be conducted by unarmed aircraft equipped with agreed sensors (optical, video, infra-red and radar) and shall be subject to annual quotas (active and passive). The entire territory of states parties shall be open to observation flights. The data recorded must be communicated to the observed party and shall be made available to other states parties.[32]

The revival of the Open Skies concept for aerial surveillance is extremely relevant. It provides yet a supplemental means of obtaining information and thus contributes to build confidence by increasing transparency of military activities. It should also be recalled that the basic principle is that each State "has complete and exclusive sovereignty over the airspace above its territory".[33] The Open Skies regime demonstrates again the flexibility which nations can exhibit regarding territorial sovereignty where security interests are deemed to require it.

The Convention on Chemical Weapons contains even more intrusive verification measures.[34] The Convention is based on each state party providing declarations on all its activities, past, present and future on chemical weapons production facilities and potential capacity. This information will be verified internationally by combination oft the data analysis and on-site inspection as appropriate to the degree of risk involved.

Concerns were voiced by the civil chemical industry over the possible misuse of information obtained during on-site inspections. An Annex on the Protection of Confidential Information defines what is to be considered confidential information. It contains rules on the employment and conduct of personnel from the Technical Secretariat, measures to protect sensitive installations in the course of on-site monitoring, and procedures to be applied in cases of breach of confidentiality rules.

Other measures have been built into the Chemical Weapons Convention to prevent abuse of monitoring procedures and challenge inspections, notably "managed access" to inspect facilities.[35] This restrictive managed access approach limits both the immediacy and the degree of access to suspect facilities and activities; however, it sets a precedent for mandatory access that could strengthen other agreements curbing proliferation.

Finally, in 1987, the Soviet Union proposed the establishment of an International Space Inspectorate (ISI).[36] Inspectors would have a right of access "for the purpose of on-site inspection, to all objects destined to be launched and stationed in space, and to their corresponding launch vehicles." This proposal has not been further discussed at the CD's *ad hoc* Committee for the Prevention of an Arms Race in Outer Space.

Lessons for Ballistic Missile Non-Proliferation

There is no doubt that the dynamics in East-West relations which emphasize "openness and transparency" have contributed to the successful conclusion of the aforementioned agreements. A close analysis of these agreements reveals several common trends.

First, all of these agreements use a combination of verification methods to ensure compliance of treaty obligations. This stems from the diversity of the obligations to be monitored. These range from efforts toward banning the testing, deployment or use of certain kinds of weapons or means of delivery, to the destruction, reduction or elimination of certain kinds of weapons, as well as the control of their transfer. With respect to chemical weapons, it concerns obligations to refrain from production, something particularly difficult to verify if the obligations overlap with civilian activities.

Second, all of them place great reliance on confidence-building measures (CBMs), notably by incorporating detailed provisions on information exchanges.

As noted by the US delegate to the group of experts appointed by the Secretary-General to carry out Comprehensive Study of Confidence-Building Measures (CBMs):

> . . . CBMs provide attractive and practical initial steps towards more ambitious approaches: no nation risks weakening its security by participating in a set of modest measures designed to improve the dissemination of certain agreed types of information relating to military matters.[37]

Third, all of them establish or refer to an organization to deal with compliance and implementation of treaty provisions. This can also be viewed as a CBM.

Proposals for Institutional Framework

The International Transfer of Missile and Other Sensitive Technologies Control efforts by the international community to control BM proliferation focus on the Missile Technology Control Regime (MTCR). Initially drafted by a group of seven states in 1987,[38] the MTCR currently has 23 Members.[39] A number of significant non-Members such as Russia, Israel and China, have promised to abide by the MTCR guidelines, while Argentina and Hungary have recently applied to become Members.

The MTCR is not a treaty but a set of guidelines to limit the conditions under which missile technology may be transferred. The guidelines do not prohibit exports but require governments to judge whether specified items meet a series of criteria before being approved for export. The MTCR also includes an

Annex listing technologies to be controlled and an informal mechanism by which the partner states can share information about potential transfers.

Though the regime has successfully slowed the overall rate of BM technology proliferation, it still faces many challenges such as: the failure of major suppliers to join the regime; the growing sophistication of production capability in potential suppliers, who also have not joined the MTCR; the increased risk of proliferation stemming from the weakness of enforceable export controls in the states from the former Soviet Union; and, above all, the fundamental inability of any supply-side control to halt proliferation.

If the MTCR is to be a more effective non-proliferation regime, it must evolve from an export control regime to a broader multilateral non-proliferation arrangement that develops and promotes international norms in the transfer and control of missile technology. In this context, France has stated that the MTCR:

> . . . should only be a stage towards a more general agreement, one that is geographically more extensive, better controlled and applicable to all. The agreement would lay down rules promoting civilian cooperation in space, while removing the dangers of the diversion of technology for a military ballistic capability. Here again, the aim would be to arrive at a situation where all states wishing to gain access to space for development purposes would cooperate in framework guaranteeing security.[40]

The negotiation of a more formal agreement would, however place greater emphasis on prohibition. Explicit prohibitions would be difficult to agree on amongst the MTCR Members, many of whom have considerable commercial interest in exporting systems and technologies for civilian space programs.

Furthermore, consideration will have to be given on how to convince key suppliers and potential suppliers to abide by the non-proliferation objectives underlying the regime, including the best way to isolate states that persist in getting a missile delivery capability for weapons of mass destruction. Based on the 1968 Non-Proliferation Treaty[41] experience, some of the more worrisome countries will undoubtedly resist MTCR limitations by pointing to the inequality of an arrangement that preserves missiles for a few states while denying them to others.

In March 1992, the United States imposed sanctions against the Russian entity GLAVKOSMOS and the Indian Space Research Organization (ISRO). The United States along with other MTCR members considered that the proposed sale violated MTCR guidelines. As stated in the US Freedom Support Act (s2532), aid can not be provided to Russia or any other republic if it violates the

MTCR.[42] Such linkages between economic and security issues can help control access to space technology.

Finally, another means of improving on the MTCR is to strengthen restrictions in terms of national legislation, which could lead to a better coordination among Member states and improve controls. It should be recalled that the guidelines depend on national legislation for their legal validity. A number of countries, notably the United States, are attempting to strengthen national control legislation.[43]

Another proposal which deals specifically with the regulation of dual-use technology was presented in 1991 in a joint Argentinean/Brazilian paper at the United Nations Disarmament Commission (UNDC).[44] This paper urges the international community to establish a set of basic rules for the international transfer of sensitive technologies. The Draft Guidelines on the International Transfer of Dual-Use Technology aims at ensuring the flow of dual purpose technology for peaceful purposes via the establishment of a mechanism which would be transparent, universal and effectively controlled through an international system of safeguards. The Draft contains no verification mechanisms.

More recently, the Legal Subcommittee (LSC) of the Committee on the Peaceful Uses of Outer Space (COPUOS) has been discussing a Working Paper tabled by a group of developing countries dealing with Principles Regarding International Cooperation in the Exploration and Utilization of Outer Space for Peaceful Purposes.[45]

Any developments along the lines proposed in the Argentinean/Brazilian paper or advocated by France will likely occur through stages. The first one might be in terms of confidence-building measures (CBMs) aimed at improving transparency. A recent example of such a CBM is the UN General Assembly Resolution on Transparency in Armaments (TIA), adopted on 9 December 1991, which calls upon the UN Secretary General to establish a register of conventional arms, including transfers, for which member states are requested to provide data according to categories established in a annex to the Resolution. The Resolution is not limited to arms transfers. Member states are also asked to inform the UN of their national arms import and export policies, as well as their legislation and administrative procedures both as regards approval of arms transfers and prevention of illicit transfers. The Resolution further invites the CD to undertake certain activities aimed at focusing attention on destabilizing weapon buildup and report back to the UN First Committee.

Of particular relevance is the request to the CD for discussion of problems associated with the transfer of high-technology with military applications. France has already elaborated on this issue by supporting the pooling and analysis of information on the national legislation, regulations and export control procedures, as a means of coping with concerns related to the transfer of dual-use technology.

This request comes at an important time in the CD's history. With the conclusion of negotiations on the CWC, the CD must choose the area for its next effort. The TIA suggests some possible directions the CD might consider.

Some states that already possess a nuclear capability have adopted a number of CBM to reduce tension with their neighbors. Argentina and Brazil have initiated on-site visits to their respective nuclear installations[46] while India and Pakistan have signed an agreement which prohibits attack against nuclear facilities.[47]

At the forty-seventh session of the General Assembly, France indicated that it was going to propose a measure to enhance confidence by making it mandatory to give advance notice of the firing of BM and rockets carrying satellites or other space objects. That notification, if adopted, would be complemented by the establishment of an international center, under UN auspices, responsible for collecting and using the data received.[48]

France elaborated its proposal in March 1993, in a Working Paper which it submitted to the *ad hoc* Committee on the Prevention of Arms Race in Outer Space of the CD. France proposed the establishment, through a new international instrument which could be negotiated at the CD, of a *regime of prior notification of launches* of space launchers and ballistic missiles, and that such a regime should be supplemented by the establishment of an International Notification Center responsible for the centralization and redistribution of collected data, so as to increase the transparency of space activities. The Center would be set up under the auspices of the UN and legally attached to it. The main function of the Center would be: to receive notification of launches of ballistic missiles and space launches transmitted to it by states parties; to receive the information transmitted by states on launches carried out; states possessing detection capabilities, are invited to communicate to the Center, on a voluntary basis, data relating to launches detected by them; and, to place the mentioned information at the disposal of the international community through a data bank.

During the past two to three years, the AH PAROS discussions have focused primarily on CBMs. The current interest in CBMs for outer space represents a first area where consensus could be reached. Based on a proposal made by the delegate of Argentina, UN General Assembly Resolution 45/55 B calls for the creation of a group of experts on CBMs for outer space.

CBMs for outer space are perceived as a means to decrease the risk of misunderstandings and ultimately conflicts arising out of incidents in space, to increase the clarity of space activities and to promote the safety and interest of all states and enhance their security.

Notes

1. US Statement to the First Committee, Political and Security Affairs, UN General Assembly, January 1957. In the same year, Canada, France and the United Kingdom suggested that a subcommittee of the Committee on Disarmament establish a technical committee to study the possibilities for an inspection system to ensure that objects sent through outer space would be used exclusively for peaceful and scientific purposes.

2. Vlasic, I.A., "The Legal Aspects of Peaceful and Non-Peaceful Uses of Outer Space" in, Jasani, B. (ed), *Peaceful and Non-Peaceful Uses of Outer Space: Problems of Definition for the Prevention of an Arms*, Taylor & Francis, London, 1991, p. 52.

3. Frye, A., "Zero Ballistic Missiles," *Foreign Policy*, Fall 1992, 2, at 14.

4. USSR Draft Treaty on General and complete Disarmament Under Strict International Control" (1962) as amended in 1964, UN Doc. DC/205, Annex 1, Section D (ENDC/2/Add.1) and UN Doc. DC/203, Annex 1, Section C (ENDC/2); and "Outline of Basic Provisions of a Treaty on General and Complete Disarmament in a Peaceful World" (1962), UN Doc. DC/203, Annex 1, Section F (ENDC/30) and UN Doc. DC/205, Annex 1, Section E and F (ENDC/30/Add.192).

5. Art. 2(4) states: "In all countries the IDO shall have its own staff, recruited internationally and in such a way as to ensure the adequate representation of all three existing groups of states..." (emphasis added).

6. Art. 43.

7. UN Doc. ENDC/PV. 37 (1962), at 28.

8. Art. 15 of the Soviet proposal provides that:

1. The launching of rockets and space devices shall be carried out exclusively for peaceful purposes.

2. The International Disarmament Organization shall exercise control over the implementation of the provisions in paragraph 1 of this article through the establishment, at the launch sites for peaceful rocket launchings of inspection teams, which shall thoroughly examine every rocket of satellite before its launching.

9. Art. 5.

10. See, Stojak, L. "International Verification Organizations for Arms Control in Outer Space," in Morris, E. (ed), *International Verification Organizations*, York University, Toronto, 1991, 125, at 126.

11. Terms used by Fedele, in Fedele, F., "Overflight by Military Aircraft in Times of Peace," (1967) 9 Jag. L. Rev. 15.

12. Freedman, L. and Schear, J.A., "International Verification Arrangements," Space Policy, February 1986, 16 at 17.

13. Treaty Between the United States of America and the Union of Soviet Socialist Republics on the Elimination of Their Intermediate-Range and Shorter-Range Missiles, signed on 8 December 1987.

14. Treaty on Conventional Forces in Europe, Paris, 19 November 1990.

15. Vienna Document of 1992 of the Negotiations on Confidence-and-Security-Building Measures convened in accordance with the Relevant Provisions of the Concluding Document of the Vienna Meeting of the Conference on Security and Co-operation in Europe, Vienna, 4 March 1992.

16. Draft Convention on the Prohibition of the Development, Production, Stockpiling and Use of Chemical Weapons and on their Destruction, CD/CW/WP.400/Rev.2, 10 August 1992.

17. Treaty on Open Skies.

18. The primary objective of the portal monitoring method is to provide continuous surveillance of the output of key production facilities. Portal perimeter monitoring may involve a series of procedures including the establishment of sensors such as X-ray devices to inspect the contents of containers passing through gates, and tag systems to detect/deter clandestine production. See, Cleminson, R.F. and Gasparini Alves, P., "Space Weapons Verification: a Brief Appraisal" in, Sur, S. (ed), *Verification of Disarmament of Limitation of Armaments: Instruments, Negotiations, Proposals*, UNIDIR, Geneva, 1992, 177, at 188.

19. 27 I.L.M. (1988), at 218.

20. Art. XII.

21. See the US-USSR bilateral "Agreement on the Establishment of Nuclear Risk Reduction Centres," CD/815 (1988).

22. Art. 3 of Protocol I.

23. Art. XV.

24. See Sur, *supra*, at 234.

25. Art. II (T) defines "reduction site" as a clearly designated location where the reduction of conventional armaments and equipment limited by the Treaty in accordance with article VIII takes place.

26. Section I, para. 1(z) of the Protocol on Inspection defines "certification site" as a clearly designated location where the certification of re-categorized

multi-purpose attack helicopters and reclassified combat-capable trainer aircraft in accordance with the Protocol on Helicopter Recategorization and the Protocol on Aircraft Reclassification takes place.

27. For a general discussion see, Macintosh, J., "International Verification Organizations: The Case of Conventional Arms Control," in, Morris, *op. cit*, 73.

28. Section I.

29. Section VIII, paras 112-142.

30. See, Morris, E., "Measures to Facilitate Transparency" in, Sur, 153.

31. Art. III.

32. Art. IX.

33. Art. 1 of the Chicago convention.

34. See, Bernauer, T., "The Projected Chemical Weapons Convention: A Guide to the Negotiations in the Conference on Disarmament," UNIDIR, United Nations, New York, 1990.

35. See Annex on Implementation and Verification, Part X.

36. Statement by E.A. Shevardnade before the Geneva Conference on Disarmament, 6 August, 1987.

37. UN Doc. A/34/416 (1979), P.58.

38. The original participants of MTCR were Canada, France, Germany, Italy, Japan, United Kingdom and the United States.

39. Australia, Austria, Belgium, Canada, Denmark, Finland, France, Germany, Greece, Iceland, Ireland, Italy, Japan, Luxembourg, the Netherlands, New Zealand, Norway, Portugal, Spain, Sweden, Switzerland, the United Kingdom and the United States.

40. Letter dated 3 June 1991 from the Representative of France Addressed to the President of the Conference on Disarmament Transmitting the Text of the Arms Control and Disarmament Plan Submitted by France on 3 June 1991" CD/1079 (1991).

41. Treaty on the Non-Proliferation of Nuclear Weapons, 729 U.N.T.S. 161 (1968).

42. See Smith, M.S., "Buying Russian Technology - -Pros and Cons for the US Programme," Reports, *Space Policy*, November 1992, 362.

43. For a discussion see, Gasparini Alves, P., *Outer Space Technologies and International Security*, UNIDIR Publ., 1991, 116-19.

44. See, "The Role of Science and Technology in the Context of International Security, Disarmament and Other Related Fields," International

Transfer of Sensitive Technology, Working Paper submitted by Argentina and Brazil, Disarmament Commission, A/CN, 10/145 (1991).

45. Working Paper presented by Argentina, Brazil, Chile, Columbia, Mexico, Nigeria, Pakistan, the Philippines, Uruguay and Venezuela.

46. See the *Agreement for the Exclusively Peaceful Use of Nuclear Energy* reproduced in CD/1117 (1991).

47. See discussion in, Gasparini-Alves, *supra*, note, at 125.

48. Official Records of the General Assembly, Forty-seventh Session, Plenary Meetings, 8th meeting, Statement by Mr. Dumas, 23 September, 1992.

9

A Flight Test Ban
as a Tool for Curbing
Ballistic Missile Proliferation

Lora Lumpe

Limitations on the testing of United States and Soviet ballistic missiles were considered throughout most of the Cold War (see Appendix A). Explicit and implicit restrictions were eventually adopted in the ABM, SALT II, INF, and START treaties. In this paper, I ask whether missile flight test limits (or a ban) are a useful tool to slow (or halt) missile proliferation, and whether such limitations are feasible in the near term. Several issues concerning the feasibility of flight test bans are considered here:

- How effective would a flight test ban be in limiting ballistic missile development?
- Could a ban be verified with a high degree of confidence?
- What complications would be introduced by flights of space launch vehicles?
- What would be the "costs" of a comprehensive flight test ban to the great powers, in terms of their force modernization plans and need for reliability testing?
- Are the United States and Russia (perhaps also Europe and China) sufficiently concerned about missiles in developing countries and each others' missiles that they would be willing to give up testing? Are countries of real concern likely to develop long-range missiles in the next several decades?

- Would a US-Russian agreement to forgo missile tests convince third world countries that they too should forgo developing ballistic missiles, given the military force inequality that would remain?

I finish by exploring the various possible configurations that flight test limitations could take.

Flight Testing

Flight testing a ballistic missile is not simply a matter of firing off a missile and watching through binoculars. The United States' two long-range test facilities (one on the east coast and one on the west) are very expensive and complex. The estimated cost of replacing the US Army's Kwajalein Atoll facility (an integral part of the Western Test Range, where all US intercontinental-range ballistic missiles (ICBMs) and ballistic missile defenses are tested), is $2 billion.[1] The fiscal year 1994 funding request for operating and modestly upgrading the facility at Kwajalein was $166.6 million (down $10.2 million from the year before), and total employment is about 3,000.[2]

The United States' Eastern Test Range, where submarine launched ballistic missiles (SLBMs) are flight tested and where intermediate-range ballistic missiles (IRBMs) were tested before they were outlawed in the 1987 Intermediate Nuclear Forces (INF) Treaty, is also very technologically advanced and expensive. The complex includes IRBM and SLBM launch pads, tracking and telemetry stations in Florida and Caribbean islands, and a down-range instrumented terminal area at Ascension Island.[3] The facility was upgraded in the mid-1980s in anticipation of the current Trident II D5 missile test program.[4]

The Soviet Union possessed similar flight testing ranges, with launch centers primarily at the Tyuratum Cosmodrome in Baikonur, Kazakhstan and secondarily at the Plesetsk Cosmodrome in Northern Russia. Appendix B contains a preliminary listing of global flight testing and space launch facilities.

The US Flight Testing Model

A ballistic missile lofts its payload to altitude and velocity in a powered boost phase, and then releases it to continue on an unpowered, unguided course. The key components of a ballistic missile are the propulsion, guidance and control systems, and the warhead or re-entry vehicle. Building a ballistic missile--perfecting and integrating these components--especially in a long or intercontinental-range ballistic missile, is a complex and daunting technological task.[5]

The basic design of all ballistic missiles was already defined in 1942, in the German V-2 missile. Nonetheless, materials, manufacturing and instrumentation used in missiles have improved greatly in the ensuing half century. Advances such as longer range through multi-staging, high-speed/low-drag reentry vehicles, multiple independently-targeted reentry vehicles (MIRVs), and successive generations of inertial guidance have all relied heavily on flight testing. The need for some flight testing in the development of any complete missile system is indisputable. As Farooq Hussain (a test ban skeptic) states: "[C]ertain problems--such a those associated with the prediction of ballistic trajectory bias, MIRV manoeuvering and warhead re-entry into the atmosphere--can only be resolved confidently by actual flight tests."[6]

The US Navy and Air Force put their missiles through an elaborate testing sequence. First, they conduct technology/component tests through supplementary flight testing (SFT) of new components on old boosters. Midgetman and Trident II components, for example, were tested aboard Minuteman ICBMs. The services rely heavily on SFT to develop reentry vehicles and guidance technology.

Next are research and development tests, carried out under idealized conditions: Engineers and technical contractors fire missiles from a launch pad (rather than silos or subs) on days when the weather conditions are most suitable. Later in the R&D process more realistic conditions are used. Typically, twenty to thirty flight tests of this sort are conducted for a new design.[7]

Then early production line models undergo a series of initial operational tests (or phase one operational tests) under more realistic launch conditions. These tests are used to estimate system reliability and accuracy. Thirty or forty flights are usually needed to achieve the level of confidence desired by nuclear war planners.

Next come so called "Demonstration and Shakedown Operational" (DASO) tests. Most of the tests in this category are of SLBMs, with test firings from each submarine; these are intended as much to test the sub and the crew as the missile. (For Air Force missiles, DASO tests follow R&D and precede initial operational tests.)

Follow-on tests (FOT--also called phase two operational tests) are carried out at lower rates over the life of the missile system: (1) to detect deterioration over time; (2) to check out modifications; (3) to maintain crew training and readiness; and (4) to maintain confidence and to display system performance for deterrent effect. The number of FOTs has been around six per year for ICBMs and more for SLBMs.[8] In addition, there is a regular program of aging and surveillance testing, using X-ray and other inspection techniques, static firing of stages, and testing of subsystems.

Third World Missile Testing

The procurement route, range and sophistication of the delivery vehicle, mission and payload all dictate particular flight testing patterns. For several reasons, developing country testing programs are not nearly as sophisticated or extensive as that of the United States.[9] A primary limiting factor is cost. A testing infrastructure is expensive,[10] and so are the missiles expended in tests. Many developing countries' missile inventories are wholly imported, and it is increasingly difficult to find resupply because of the emerging norm against missile exports. A meaningful test program could easily deplete the limited missile supply of a developing country.

Second, given that the vast majority of developing country ballistic missile systems have been imported, flight testing is less necessary. Thirteen countries have imported the Soviet Scud-B.[11] The Scud is a simple, proven design, based originally on the V-2. It does not require tight tolerances in its manufacture and handling and, therefore, perhaps a purchasing country could deploy it with little or no testing.[12] Similarly, Saudi Arabia purchased an estimated 50 CSS-2s from China. No operational flight tests of this 2,400 km range missile from Saudi soil have been reported.

In addition, most developing country missiles are of short range and conventionally armed. Possessor countries have used them as deterrents, or as counter-city weapons of terror. Obviously, neither mission calls for the high degree of accuracy and reliability needed for counter-silo nuclear weapons.

Because of the paucity of testing by newly ballistic-missile-capable countries, some analysts have asserted that testing restrictions would be of little utility in curbing missile proliferation.[13] However, the third world missile development or upgrade programs of greatest concern--those aimed at achieving accurate inertial guidance, solid fuel and multi-staging--must flight test. And, in fact, those few developing countries that are pursuing long range missile or space launch capabilities (Brazil, India and Israel) have serious, methodical flight testing programs, albeit with fewer flights and at less cost than for the superpower programs.

For example, the Indian military has tested its Prithvi[14] missile twelve times in the past four years (see Appendix C). The first test occurred on 25 February 1988 at the Indian Space Research Organization's SHAR Centre (on Sriharikota Island). After five more tests there, Prithvi has since been tested from the military's Chandipur interim test range in Orissa, most recently in late November 1993.[15] The Indian military will likely deploy Prithvi during 1994. India is putting its longer range ballistic missile, the Agni, and its space launch vehicles through a similar steady progression of tests.

Israel has the most highly developed defense industry in the Middle East and the most advanced military missile production capability outside of the former Soviet Union, United States, France and China. During the 1970s-1980s Israel developed an improved version of its imported Jericho missile, dubbed the Jericho II. (Many reports claim Israel is developing two separate missile systems, the Jericho II with a 800 km range and the Jericho IIB with an extended 1,300 km range.) The missile has successfully flown several times (see Appendix C). Most recently, on 14 September 1989, Israel fired a Jericho II 1,300 km into the Mediterranean. In addition to tests in Israel, two long-range tests of the Jericho II are believed to have occurred in South Africa during 1989-1990.[16] About 50 Jericho II missiles are believed to have been deployed.[17] Israel has also twice tested successfully an indigenously developed space launch vehicle, the Shavit.

As the Indian and Israeli military establishments know, zero flight testing of a missile under development will result in zero confidence that the system is functional. Moreover, achieving an acceptable degree of confidence in the reliability of a system, and characterizing its accuracy and performance under varying conditions, requires operational flight testing, the amount of which varies with the amount of information, and statistical confidence in that information, one desires to have.

Although media reports often refer to the "improved accuracy" of third world missiles, without a significant and highly visible testing program, such claims must be treated with skepticism. The measure of accuracy, circular error probable (CEP--the radius of a circle within which half of missiles launched at a target point are expected to impact), cannot be determined by a single test; CEP can only be estimated by firing a substantial number of missiles at predetermined aim points. Accuracy can be compromised by subtle imperfections in machining, calibration or system engineering, and most developing countries do not have or do not produce missiles in quantities sufficient to support testing at the rates required to assess progress in the refinement of guidance systems, or even to iron out all the bugs and glitches that may cause catastrophic failure.[18]

The absence of testing, therefore may be considered *prima facie* evidence that alleged missile capabilities are non-existent. Developing country missile programs have been exaggerated often for political reasons--both by the alleged proliferators (for reasons of deterrence or prestige) and by developed countries (to justify certain military programs and to support arms sales to allies claimed to be threatened). A clear distinction must be made between the "real" developing country missile programs (like those in Israel and India) and those chimeric programs which appear to lack rigorous (or in some cases apparently any) testing. Perhaps most notable in this regard are exaggerated (or at least unproven) claims of North Korean ballistic missile prowess. The Pyongyang government imported Scud missiles from Egypt in the mid-1970s and began producing its own version of the 300 km missile in 1987. In the late 1980s and early 1990s, a number of sources reported that North Korea was refining the Scud to increase its range to 600 km, and to improve its guidance.[19] According to former Director of Central Intelligence, James Woolsey, North Korea "is developing and actively marketing a new, 1,000 kilometer-range missile,"[20] called the Nodong I. The missile appears to have been successfully flight tested once (in May 1993 to a range of 400-500 km) and unsuccessfully tested once (see Appendix C). As it is allegedly a further modification of the Scud,[21] few tests might be needed, but more than one flight test would be expected for a serious weapons program. Various press reports claim that the system is either in late development and will be deployed and exported to other countries shortly or that it already has. (Following on the heels of the May 1990 test launch, press reports claimed that North Korea intended to extend the range of this missile to 1,300 km.[22]) At best, the functionality (let alone accuracy and reliability) of North Korean-made missiles is uncertain. According to one report, Scud missiles manufactured by North Korea and shipped to Iran in the early 1990s were inoperable, and Iran returned the missiles.[23]

Similarly, allegations of missile production by several countries in the Middle East are not supported by available information on flight tests. Whether this is because tests are not occurring, are not being observed or not being reported publicly is difficult to determine. What is clear, is that in the past few years the US military has diverted increasing intelligence assets to cover developing countries and regions considered dangerous.[24]

Alternatives to Testing

If flight testing were limited or prevented by some control regime, and if suppliers refused to transfer full-up ballistic missile systems, then could a developing country--or even a developed country--achieve an operational ballistic missile capability?

The number of flight tests needed to develop and obtain confidence in a ballistic missile is decreasing over time. This decline is due to the accumulation of knowledge and the development of improved alternative techniques for evaluating missile systems. Better instrumentation and analysis methods are applied to static firings of the rocket motors; and more sophisticated simulations of launch and flight are applied to workouts of the guidance and control systems. In addition, continually increasing computational capabilities streamline the development process by aiding design, development and evaluation.

Political considerations have also driven down the numbers and restricted the possibilities for testing. For example, overflights of the continental United States are problematic; and a static testing procedure called Simulated Electronic Launch of Minuteman was introduced to compensate partially for the lack of flight tests fired from active-duty silos.[25]

Intangible factors such as morale (in turn, affected by the threat perception) of the missile work force, organizational structure and even culture may affect the number of tests needed. A common cause of failure in complex technical projects is poor coordination and communication between teams which produce different subsystems that must work together. A better-educated and better-equipped work force may also be better able to circumvent difficult technical problems and minimize testing.

The transfer of knowledge from non-military space activities to military missiles may also reduce flight testing requirements. But it is easy to overstate the adaptability of civilian space technology to military requirements. Farooq Hussain states that, "The development of very reliable launchers, for both satellites and manned spacecraft, with a minimum number of flight tests has always been the philosophy of NASA. The United States' aerospace industry has now learned the methods by which very high reliability can be attained essentially without a flight-test requirement or with only a nominal one."[26] The trick, according to Hussain, is to "introduce a large amount of redundancy in back-up systems." However, this "very high reliability" is purchased only at a cost that would be prohibitive for most military systems. Redundancy means dead weight, and a multiplication of system cost and complexity. A typical commercial launch is carried out only under the best weather conditions, and only after the rocket has been examined by an army of technicians, and a committee of engineers has given the word "Go." Moreover, the continuing series of unmanned launches, using rockets with long histories such as Atlas, Titan, and Delta, is itself a *de facto* testing program from which information is obtained to improve the rockets and their operational use. Major upgrades of these systems have been accompanied by an expensive and embarrassing series of failures. Nor has the manned space program been without its disasters and learning curves. After the tragedy of Apollo 1, a series of unmanned launches of the Saturn vehicles was carried out before manned flight resumed; and even the space shuttle, with all its redundancy and sensors monitoring every component, has experienced one catastrophe and many mishaps.

Robert Sherman has noted, "The history of missile development is replete with examples of new missiles and new technologies which performed well in computer simulation and ground testing, but which revealed unpredicted--and probably unpredictable--fatal defects in flight testing."[27] Of the eight new strategic missiles first tested in the 1980s (MX, Trident II, Pershing II, SS-24, SS-25, SS-N-20, SS-N-23 and an SS-18 follow-on), all but two failed their first flight tests. The MX missile's inertial guidance system performed "brilliantly" in early development tests, but its accuracy fell off when the production team took over production of the missiles from the development team, according to Sherman.

A recently revealed report on Soviet ICBM development demonstrates the requirement for extensive static testing *and* operational flight testing. Apparently the Soviets had persistent difficulties with hydro-thermodynamic instabilities in their liquid-fuel engines. Production-model engines that worked most of the time would unpredictably exhibit these instabilities. Sometimes the engines failed catastrophically. No systematic differences between the engines that worked and those that failed could be detected, and despite the efforts of hundreds of scientists, the effects could not be reproduced in the lab. *Ad hoc* solutions were sometimes found, but the scientists could not obtain any systematic control over the phenomena. According to the author of the report, a former Soviet rocket engineer, this problem explains why the Soviets relied for so long on clusters of small, inefficient engines on their booster rockets, instead of moving to the larger, more efficient engines used by the United States. Their solution was to mount sensors in the engines that would shut them down at the first sign of trouble, then the rocket would fly on using the remaining engines.

This example clearly illustrates the kinds of difficulties that may arise unexpectedly in the development of missile systems, and may go undetected in the absence of a rigorous testing program. As it turned out, the clustering fix masked a systematic flaw. During the winter of 1965-6, the Strategic Rocket Forces undertook a standard test of a deployed strategic missile. The nuclear warhead was replaced with a dummy, and the missile was transported to a space facility for the launch toward the Pacific. The operational missile, which had been produced serially in the thousands, exploded on the launch pad. Hydrodynamic instabilities in the fuel feed lines caused the explosion. The oscillations turned out to be associated with a narrow range of air temperatures, around -30 degrees C. The engine had passed ignition tests at 40, 30, 20, 10, 0, -10, -20 and -40 degrees C, but had never been tested at -30.[28]

In summary, if no flight testing is permitted, then every new weapon or component risks catastrophic failure with a high probability.

Incentives for Other Third World Countries

The present response by the developed world to the spread of missile technology is an export control/proliferation management regime, combined with a "technical fix" in the form of anti-missile systems. The Missile Technology Control Regime (MTCR), initiated in 1987 with seven members, has grown to include 25, mostly Western industrialized countries.[29] MTCR members and adherents pledge to abide by common export guidelines on missile-relevant technologies and missiles themselves. Although a factsheet on the MTCR issued by the US government in 1987 said that the guidelines "are not designed to impede national space programs or international cooperation in such programs as long as such programs could not contribute to nuclear weapons delivery systems,"[30] the regime seems to have acquired the goal of doing precisely that. By definition, any space launch vehicle could contribute to a ballistic missile that conceivably could deliver a nuclear payload.

At the same time, many countries of the North, alarmed about the perceived spread of missile capabilities, are developing or purchasing anti-missile systems. The US government plans to spend $3.2 billion in Fiscal Year 1995 to develop ballistic missile defenses, and the United States and Russia are aggressively marketing their tactical anti-missile missiles to countries in the Middle East and East Asia.

Unfortunately, the MTCR and missile defenses do not address the security concerns of the "proliferators." These policies have therefore failed to eliminate the demand for missiles, borne of regional political tension and local arms races.

Janne Nolan has argued that even if the big military powers agreed to forgo missile flight testing in order to curb missile proliferation, developing countries would not find this offer all that compelling, given the disparities that would remain in the size and capabilities of military arsenals.[31] Yet developing countries would gain some palpable security benefits through a flight test ban. Superpowers like to think of ballistic missiles primarily as deterrents. But in the developing world, missiles have been used recently and extensively against cities, mainly as weapons of terror and attrition.[32] A testing ban would immediately improve the security environment of many countries by halting costly and destabilizing regional missile races. If the ban extended down to missiles with a range of 100 km/500 kg payload, then the benefits would be even greater, as deployed short-range (yet often "strategic") systems would be "rusted out" over time.

Such a regime would not equalize the global military imbalance. In particular, the United States and Russia would retain massive conventional superiority as well as massive nuclear capabilities. But in time, these arsenals would also dwindle through lack of operational readiness training, gradual loss of confidence, and (hopefully) eventual strategic irrelevance.

The entire world would benefit by decreasing the chance of accidental or intentional nuclear war. A flight test ban should also alleviate the perceived need for anti-missile systems, lessening global tensions and freeing up vast resources that would be spent to develop and deploy such systems.

Certain developing countries would also be relieved of anxiety about the United States and Russia re-targeting ICBMs on them. A Pentagon report leaked to the press in January 1992 suggested that in the post-Cold War world, "every reasonable adversary"--some presumably in the developing world--should be targeted with nuclear and non-nuclear strategic weapons.[33] As part of its recent "counter-proliferation" initiative, the US Department of Defense is reportedly considering fitting some Trident II D5 missiles with small nuclear weapons,[34] as well as with conventional warheads. Rear Admiral Thomas Ryan, the director of the US Navy's submarine warfare division, argues that the latter is needed as a credible deterrent against certain developing countries.[35] A long-range, kinetic energy penetrator is intended to destroy underground command and communication bunkers of potential (third world) adversaries. This wild plan appears to be driven by the search for a new mission for the D5, which was to have been targeted primarily on the hardened SS-18 silos, now scheduled to be eliminated under START II.[36] The bunker-busting mission would require accuracy of 5-7 meters, which could not be achieved without testing;[37] on November 18, 1993 the Navy conducted a classified test of a D5 missile equipped with at least two conventional warheads from a Trident submarine off the Florida coast.[38]

US Navy officials are touting a conventional SLBM as an "anti-proliferation weapon." But it is likely that this current (and disturbingly recurrent) talk of converting ICBMs or SLBMs to engage third world targets from intercontinental range will motivate developing countries to pursue their own long range missile development. In addition, the development of ultra-high accuracy needed for conventional SLBMs could destabilize the US-Russian nuclear relationship and re-energize the qualitative nuclear arms race.

Great Power Issues

Undoubtedly, flight testing restrictions would hamper and even make impossible the spread of long range missile capability. But the major military powers may be unwilling to forgo testing to achieve this end. Nolan states that a flight test ban (FTB) "has never been considered seriously by [the five declared nuclear states'] governments,"[39] and that "the notion that the superpowers and the NATO allies would abandon missile flight testing in the hopes of persuading third world countries to follow suit lacks credibility."[40] In this section, therefore, I analyze the crucial issue of the interests of current ICBM powers.

Russia

Some Russian legislators perceive START II to be unfavorable to Moscow. Others worry that the costs of the treaty are too high. They estimate that it will cost Russia $5 billion to comply with START II and to reconfigure its forces, including producing new ICBMs not prohibited by the treaty.[41] The SS-25 is the only ICBM currently produced in Russia.[42] US intelligence sources reportedly expect that Russia will flight test and deploy one or two follow-ons to the SS-25 and to the SS-N-20 SLBM sometime later this decade.[43] The Russian press reported in early April 1993 that the Moscow Thermo-Engineering Institute and the Dnepropetrovsk "Yuzhnoye" Science and Production Association are developing a new "multipurpose" ICBM.[44]

According to CIA Director Woolsey, "Russia's willingness to fulfill START II requirements will depend, in part, on its ability to modernize its remaining forces to make them viable into the next century and to ensure that it remains a strategic super-power."[45] On the other hand, some $1.2 billion in so-called "Nunn-Lugar" monies are being provided to assist in the denuclearization of the country, but continued strategic modernization is undercutting political support in the US Congress for more assistance.

A converted SS-25 underwent an initial test launch as a space launch vehicle from the Plesetsk Cosmodrome on March 25, 1993. The commercial launcher is known as the `START I'.[46] In addition, a Washington, DC-based group of investors, headed by former Chairman of the Joint Chiefs Admiral Thomas Moorer, has also contracted with a Russian company to develop a mobile commercial satellite launcher. Their so-called `Surf' vehicle would incorporate elements from deactivated SS-N-23 and SS-N-20 SLBMs and would be launched from mobile floating platforms.[47] Several other planned conversions are underway.[48] Such schemes would complicate a ballistic missile flight test ban, but they do not present insurmountable obstacles.

United States

A flight test ban would preclude the continuation by the United States and Russia of the race for exotic first strike weapons,[49] such as high-accuracy usable capabilities, defenses, depressed trajectory[50]/short time of flight weapons, maneuvering reentry vehicles (MaRVs) and precision-guided RVs.

A ban would also erode the reliability and confidence essential to first-strike planning.[51] As Robert Sherman has argued, "statistical analysis can demonstrate that deterrence and stability are highest when strategic missiles on both sides are `semi-rusted'--that is, when only about 30-70 percent can be expected to work properly...Less-than-perfect reliability discourages aggression more than it impairs deterrence."[52] In spite of this common-sense logic, supporters of an FTB will collide head-on (as do supporters of a nuclear test ban) with proponents of weapons safety and reliability testing. In response to Sherman, Walter B. Slocombe argued that reliability problems would "breed concerns about massive undetected problems with the force, which would foster pressures for early use or, more probably, for abandoning the flight test ban... Implicitly, a ballistic missile flight test ban reflects the view...that nuclear weapons modernization itself is the chief source of danger."[53]   It is difficult to know what Slocombe means by "pressures for early use," but one need not be wedded to the view that modernization is *the* wellspring of crisis instability to agree that it can be *a* source of danger.   Moreover, it is hard to envisage any arms control agreement which does not pose the "danger" that it will someday be broken.

With respect to operational reliability, a chart at the recent authorization hearing for the Trident II program noted that "Accuracy, reliability and safety can only be *objectively* evaluated, verified and predicted by flight test...The ability to *detect* and *correct* an unacceptable degradation in accuracy, reliability or safety is required."[54]   This statement confuses the information that is obtained in early development and evaluative test series with that obtained later in continuing operational tests.   The latter are not conducted at a rate sufficient to detect the changes that might be expected in accuracy or in reliability rates.   A sudden and unexpected degradation probably would remain undetected for a considerable period if testing were relied upon to reveal it.   But in fact, the primary guard against such deterioration is inspection and non-flight testing of the operational missiles.[55]

Farooq Hussain argues that an FTB would impede many "*desirable* technological improvements to existing systems which contribute to increased survivability."[56]   Contributions to survivability, either against first-strike attack or against ballistic missile defense systems, do contribute in principle to stability. But only improvements in survivability against defenses depend critically on flight testing.[57]   Thus if defenses remain limited under the ABM Treaty, we need not fear an erosion of crisis stability due to an FTB.

Finally, some in the US defense establishment maintain that an end to testing would be detrimental to US security, in that it would lead to greater uncertainty about other countries' missile capabilities.   During flight tests, missiles transmit a stream of electronic data on the missile's performance to monitors on the ground. Interception of this data, called telemetry, reveals the capabilities of the missile. Denial of telemetry and of visual information through a testing ban, it is said, would lead to worst case threat analyses, which would spur arms races.[58] However, worst case analyses prevail anyway, without testing restrictions.   The trade off here is one of curtailing missile development through a negotiated test ban, or allowing development to continue simply so that it can be monitored.

There are strong political interests in the United States committed to continuing ballistic missile testing.   Continued testing is driven by: (1) bureaucratic self-maintenance of the US `missile-lab complex'; (2) the market provided by testing for the US aerospace industry; (3) the quest to develop missile defenses; and (4) the need to maintain superiority over British and French nuclear forces (to justify US leadership of the NATO alliance) and over Russian and Chinese countervailing forces.[59]   Consequently, the United States has conducted 20 to 30 ICBM launches annually in recent years. Similar pressures-- especially pressures to maintain missile industry jobs--are at play in Russia.

The Trident II (D5), which began launch pad flight testing in 1987, is currently the only ICBM in production in the United States.   After 19 developmental tests and nine "performance evaluation missile" tests, the missiles are currently being tested in each of the D5 capable nuclear submarines (SSBN). On 20 August 1993, the eleventh Trident II DASO test took place, this one launched from the USS *Nebraska*.[60] The Navy has slated 35 missiles for DASO tests.[61] The fiscal year 1994 US defense budget request reflects the reduction in planned procurement of D5 missiles (down from 779 to 428) and reduced operational testing plans.[62] Because of the success of the Trident II test program to date, the Navy ended the CINC Evaluation Test (CET) program early and initiated a reduced follow-on CET.   The program provides data to evaluate continually the performance of the system through the design service life (at least 30 years) of the D5 capable Trident subs. The Navy has slated 138 Trident II missiles for CET and follow-on CET.[63]

In addition, the US Air Force is urging guidance and propulsion system replacement programs for the Minuteman III, which, after the year 2003, will be the only remaining land-based ICBMs in the United States.[64]

France

In February 1993 M-4 missiles[65] were retrofitted onto the last of five French strategic nuclear submarines. Each of the five submarines can launch 16 M-4 SLBMs, each of which carries up to six 150kt warheads.[66] France plans to have operational a new generation submarine (Triomphant class) by mid 1995. These will carry 16 M-45 missiles each.   Unlike the M-4, M-45 missiles carry electronic counter-measures and penetration aids.   France plans, in turn, to replace the M-45 missiles with 8,000-9,000 km range M-5 missiles in the 21st century.[67] France has no ICBMs, but has developed short and intermediate range land-based ballistic missiles. France tests its IRBMs and SLBMs into the North Atlantic, with a down-range tracking station on Saint Maria Island in the Azores.[68]

United Kingdom

Britain neither builds nor tests its own ICBMs, but rather procures Trident II (D5) missiles from the United States and tests its missiles and subs on the US Eastern Test Range. The United Kingdom is reportedly slated to purchase 67 D5 missiles (including test missiles and spares) for four Trident class subs during the next 30 years.[69]

China

China is developing new ICBMs and SLBMs. Deployment of the DF-31 (ICBM) and the JL-2 (SLBM) are scheduled for mid- to late 1990s, and an even more ambitious 12,000 km range mobile missile, the DF-41 is on the drawing board.[70] In addition, China is working on improved guidance and MIRVs.[71] China has flight tested missiles into the Pacific, to near the Solomon Islands, and into the Yellow Sea and Indian Ocean.

Possible Controls

Comprehensive Flight Test Ban (CFTB)

A comprehensive flight test ban would be more effective in impeding missile development than would any partial measure, and it would be more attractive politically to developing country missile aspirants. A CFTB would level the playing field between them and the great powers in terms of permitted and non-permitted missile-related activities. Furthermore, a CFTB would be far easier to verify than existing arms control undertakings (like, for example, the Chemical Weapons Convention and even the Nuclear Non-Proliferation Treaty). The development of a new ballistic missile system cannot be kept secret. As then CIA Director William Webster acknowledged in May 1989, "The status of missile development programs is less difficult to track than nuclear weapons development. New missile systems must be tested thoroughly and in the open..."[72] Flight testing is unavoidably observable--and becomes more easily observable the longer the range of the missile. US early warning satellites can determine reliably whether missiles are or are not being flight tested.[73]

The use of possible ballistic missile components in space launch vehicles to circumvent a flight test ban would complicate such a regime.[74] However, even if some component testing could not be prevented, the lack of complete system tests would result in low confidence in missile reliability. As Sherman notes: "War planners are aware of the numerous instances in which components worked perfectly by themselves but, when flight tested together, revealed disastrous incompatibilities that otherwise would have been undiscovered."[75]

Moreover, interception of non-encrypted telemetry signals would expose military-related upgrades on ostensible space launcher flights. Non-encryption of telemetry should be a staple of any FTB regime. Violation of this principle might provide early indication of intention to break out of a missile flight ban treaty.

Undoubtedly, a CFTB would involve tradeoffs between arms control effectiveness and non-interference with space activities. If too lax, flight testing restrictions might be ineffectual, allowing the transfer of improvements from the civilian sector to the military. If too severe, restrictions might impede civil space programs. To build the strongest possible wall between ballistic missile tests and space flights, Robert Sherman has suggested the following guideposts at each stage of flight:[76]

Reentry. High-speed reentry, radar-emitting reentry vehicles and terminal maneuvers could be prohibited. Ballistic missile reentry vehicles approach or impact the earth at many times the speed of sound. Accuracy would diminish if they were slower and spent more time in the atmosphere. High-speed reentry is not used in space programs, however, because "it would be bad for reusable payloads and worse for astronaut morale." In addition, legally permissible re-entry angles could be defined to distinguish between legitimate space booster rockets and ballistic missiles and between satellite/shuttle/spacecraft re-entries and weapons payload re-entry vehicles.[77]

Warhead separation phase. The weights and profiles of existing reentry vehicles could be catalogued, and the release of objects sharing the weight and velocity change of missile reentry vehicles could then be banned.

Boost stage. Each party to the flight test ban would list the length, diameter and total impulse of every missile boost stage it deploys; flights of these devices could be prohibited. Where boosters are identical to space launch vehicles, the space boosters must be displayed for inspection, counting and tagging. Tagged boosters would be granted an exemption from the test ban, provided they were not flown on a missile trajectory. When the tagged boosters were expended, all new boosters would have to be verifiably different.

In addition, all US ICBMs and the more modern of former Soviet ICBMs use solid-fuel rocket engines. Solid propellants are more stable and storable than are liquid fuels, making them more militarily useful. Through such a cataloguing and tagging system as Sherman proposes, new space launch vehicles could be required to utilize non-storable liquid fuel engines. Adherence could be verified during flight by infrared sensors, which can determine the chemical composition of rocket propellant by its thermal signature.[78]

Guidance systems. Ensuring that guidance being tested on a space shuttle or space launch vehicle is not intended for an ICBM is the most formidable challenge. To deal with this, Sherman recommends internal inspection of missiles and space vehicles.

Partial Test Ban

Several possible configurations for regional bans or other partial FTB measures exist.

Numerical testing ceilings. A regime could be negotiated that permitted only *X* number of tests per year. An annual quota of perhaps five or six flight tests (similar to a proposal by President Carter in 1977) would allow the major military powers to maintain confidence in the reliability of their arsenals while slowing modernization and development. Such a ceiling was proposed by Sidney Drell and Theodore Ralston in 1985. They calculated that a 50 percent cut in tests (from the average twelve per year to six) would cause a major delay in achieving confidence in accuracy enhancements.[79]

Hussain notes that the Soviet development philosophy resulted in flight testing more often than did the US approach,[80] a fact which could make a numerical ceiling more difficult to negotiate. However, with the radical improvement in relations between the former Soviet Union and the United States, missile modernization is already slowing. On the other hand, since most third world countries undertake only a small number of tests annually, a test ceiling would do little to prevent continued development and proliferation.

The ceiling could be designed to allow only operational/reliability tests, preventing testing of innovations. Verification of a ban on development tests would be complicated, as it would have to ensure that incremental improvements (for example, guidance improvements) were not being flight tested clandestinely on surrogate missile launchers. The broadcast of un-encrypted telemetry would help verify against such cheating.

A delivery range-delimited ban. The first question is: what range? Many of the current developing country missiles of concern are short range systems (100-300 km). Missiles of similar range in the US arsenal (like the Army Tactical Missile System) are considered tactical battlefield missiles. The impact on Russian short range systems must be factored in also to their likely acceptance or non-acceptance of such a regime.[81] An agreement to ban tests of such short range systems would be more difficult (but not impossible) to verify.

The United States and Soviet Union agreed in 1987 to renounce their intermediate-range ballistic missiles, those with a range from 500-5,500 km. The INF Treaty also prohibits the signatories from flight testing systems of this class. Former ACDA officials Kenneth Adelman and Kathleen Bailey have promoted the idea of internationalizing the INF Treaty as a possible approach to curtailing third world missile proliferation.[82] However, nearly all of the systems currently deployed by developing countries would fall below the 500-5,000 km range covered by INF. The ubiquitous Scud-B, for example, would not be included, nor would the SS-21, Lance, or Jericho I systems. In the Middle East, only the Israeli Jericho II, the Saudi CSS-2, the Indian Agni (under development), North Korean missiles under development and Iraqi missiles (which are now being destroyed anyway) would be covered. A ban on testing systems of these delivery ranges would be a meaningful step, but a regime that left their adversaries' missiles in place, and would not permit testing of their own systems, would likely be unacceptable to the Israelis and Saudis.

Such a regime would also leave open the possibility, however slim, of third world countries leapfrogging ahead to missiles with delivery ranges above the 5,000 km INF ceiling, approaching ranges that could strike the continental United States.[83]   Further, since some developing-country ballistic missile programs are driven partly by arms races or tension with ICBM-possessing countries (for example, India's concern with China), such a non-inclusive regime might lack support.

Since there is a direct relationship between missile payload and range, the possibility of downloading payload to achieve a greater than permissible range must be factored also in to any range-delimited test ban.

Flight test free zone (FTFZ).   A negotiated FTFZ already exists, in the demilitarized Antarctic, and several other regions (Latin America, most of Africa) are *de facto* FTFZs.   In the 1970s the Campaign for a Nuclear Free and Independent Pacific attempted to incorporate an FTFZ into the 1986 South Pacific Nuclear-Free Zone Treaty.[84]   Australia quashed this effort.

A geographically-delimited approach has several advantages for curbing missile proliferation.   In particular, it is not dependent on gaining the agreement of all, or nearly all, of the states currently deploying or developing ballistic missiles.[85]   Only some subset of this group would be required to go along.   A principal difficulty with this approach, however, is that many of the regional arms races overlap each other.   For example, a regional FTFZ to include Pakistan and India would probably need to include China; and a Middle East FTFZ might spill over to include Pakistan and India.

The lack of a negotiating history between many of the regional adversaries engaged in missile races may make global or regional FTFZs premature.   It might make more sense at the outset to engage regional adversaries in confidence-building exercises and to foster regional peace processes and reconciliation.   Conversely, a missile flight test ban is one of the more meaningful and readily verifiable arms control measures imaginable.   It is conceivable that regional adversaries--especially those that have alliance partners outside the region--might find a regional FTFZ to be a productive diplomatic strategy to build confidence within the region.

In the Middle East, some tentative steps toward a regional FTFZ were broached by the United States through separate talks with the Egyptian and Israeli governments in the late 1980s.   Under discussion, reportedly, were small confidence building steps such as advance notice of missile flight tests and possibly "no first use" pledges that could lay the groundwork for farther-reaching steps in the future.[86] In his post-Gulf War Middle East arms control initiative of 29 May 1991, President Bush called for a halt to further acquisition, production and testing of ballistic missiles of any range by states in the region, leading to "the ultimate elimination of such missiles from their arsenals."[87]

This proposal would appear to be in the interests of all countries of the region. The citizens of Israel, Iran, Iraq and Saudi Arabia have all been threatened and attacked by ballistic missiles in the past five years. Moreover, an FTFZ would meet both the Israeli and the Arab states' arms control interests. In general, Israel advocates limits on conventional arms transfers to the region, while the Arab states prefer to deal with unconventional weapons first, conventional arms later. Because of their historical use as delivery vehicles for nuclear payloads, and because of their relationship to conventional air force capabilities, ballistic missiles fall into a grey area. Thus, missile disarmament might be acceptable to both sides as a first step. Indeed, Egyptian President Hosni Mubarak has vigorously endorsed a plan for a zone free of weapons of mass destruction in the Middle East, which calls on all Middle Eastern countries to announce their commitment to "deal effectively and honestly with matters involving the delivery systems of various weapons of mass destruction."[88]

A regional testing regime that permitted flight testing only of already-deployed systems would probably be unacceptable to regional actors that felt at a disadvantage (that is, did not then deploy missiles). Conversely, a regime that allowed all countries in a region to develop missiles up to the longest range missile deployed in the region (for example, to the range of the CSS-2 or Jericho II/Shavit in the Mideast), thereby permitting development of long-range missiles by non-allies like Iran or Libya, would be unacceptable to the United States. A total, regional missile flight test ban seems the most likely to be accepted. A central question is whether regional flight test bans could be agreed without a global ban, or at least without superpower participation in the form, for example, of no-first use guarantees.

An FTFZ approach might have to account also for tests conducted by the parties on the territory of another country, outside the region. This situation has occurred quite often in third world missile development (for example, Israel apparently tested in South Africa; Iraq reportedly tested missiles in Mauritania; and Iran reportedly has recently prepared to flight test missiles in Sudan).

An RV test ban or ban on testing new MIRVs. Although this ban would be useful in the great power context (including China), it would not apply to most developing countries. However, it might be desirable to lock all potential long range missile countries into such a ban preemptively to preclude MIRV development.

START II bans, among other things, flight-testing MIRVed ICBMs after 1 January 2003.   Director of Central Intelligence James Woolsey told the US Senate in summer 1993, "We will be able to monitor the ban on MIRVed ICBMs...both by tracking the elimination of launchers for MIRVed ICBMs and by analyzing the data from flight tests of new missiles."[89]

Ban certain flight trajectories. The most obvious candidate here would be a ban on flight testing missiles (especially SLBMs) on a depressed trajectory (DT).[90] This step would be useful in the great power context. But it is irrelevant in the near future to developing country missile proliferation.   Since such a trajectory would have no overlap with space flights, it would be easy to verify a DT ban.

En route to achieving flight limitation regimes, several confidence inspiring measures could be undertaken.   During the Cold War, the United States and Soviet Union pre-notified each other of missile tests and broadcast warnings to mariners of the expected area of missile impact (see Appendix A).   These practices could be expanded to include any country firing missiles into international waters or overflying another country. States in a given region could also pledge that missiles will be tested on non-proactive flight paths, away from adversaries' land mass or other assets.

Conclusions

With the Cold War over, security analysts have identified ballistic missile and nuclear weapons proliferation as the leading threat facing the United States now and in the coming years. The US defense establishment perceives the threat from third world missile development programs to be serious enough to warrant an outlay of several billions of dollars per year to develop technical countermeasures and 'non-proliferation' programs.

In reality, the scale of the missile proliferation threat has sometimes been exaggerated--by proponents of missile defenses and by `proliferators' themselves. Many of the countries often cited as being a source of concern actually have very limited indigenous programs. Many undertake little or no flight testing, which may be less necessary for imported arsenals of proven and primitive systems like the Soviet-manufactured Scud-B. But missile flight testing is essential to achieve any degree of confidence that a ballistic missile system under development will work as intended. Once the system has been tested adequately, operational reliability can be assured to some degree with methods other than flight testing. Certainly, a global and total FTB would freeze existing ballistic missile developments, and gradually erode those holdings over time. In order to ensure that clandestine development of ballistic missiles was not occurring under the guise of space launcher tests, some special provisions would have to be made. However, an FTB would be more easily verified through satellite and aircraft reconnaissance than any other arms control agreement imaginable.

A few third world countries are steadily developing substantive space launch vehicles and long-range ballistic missiles. India and Israel (and to a lesser stage of development, Brazil) have missile development programs, demonstrable through serial flight testing. Israel and India are not particularly politically worrisome to the United States, but both have nuclear weapons. Eventual development and deployment of nuclear tipped (or possibly nuclear tipped) ICBMs by them would have far-reaching implications.

In addition, China continues to develop more advanced strategic nuclear weapons. In the 1990s, China is expected to deploy three new ICBMs/SLBMs, as well as its first MIRVed missiles. This eventuality would also be globally destabilizing. The perilous political fate of pro-Western politicians in Russia increases the desirability of a ballistic missile flight test ban.

Agreement to an FTB in the near term would also demonstrate the commitment to nuclear arms reduction which the superpowers pledged as an inducement to countries to sign on to the Nuclear Non-Proliferation Treaty in 1969. In 1995 the adherents to the NPT will decide whether and for how long to extend the Treaty. Such a ban would impede US Navy and Air Force plans, as well as those of the other declared nuclear powers. The main question relevant to the establishment of a global FTB regime is whether the United States believes that non-proliferation benefits accrued from an end to testing outweigh the bureaucratic imperatives, and psychological needs to continue testing.

Appendix A: Flight Test Control or Notification Measures Explored or Undertaken

Early 1960s: Jerome Wiesner, science advisor to President Kennedy, considers the desirability of a US proposal to the Soviet Union to ban missile flight tests. NASA opposes the idea, not wanting interference with space launcher programs. The US Arms Control and Disarmament Agency and the State Department authorize studies to consider the possibility of preventing the further development of ballistic missiles through flight testing restrictions. A dozen or so constraints are considered, but NASA rejects them all.[91]

1961: The Antarctic Treaty prohibits the testing of any weapon system in Antarctica.

1971: Agreement on Measures to Reduce the Risk of Outbreak of Nuclear War Between the United States of America and the Union of Soviet Socialist Republics, Article 4 mandates that "Each party undertakes to notify the other

Party in advance of any planned missile launches if such launches will extend beyond its national territory in the direction of the other Party."

1972: Agreement Between the Government of the United States of America and the Government of the Union of Soviet Socialist Republics on the Prevention of Incidents on and Over the High Seas, Article VI establishes broadcasting by radio a warning to mariners and other shipping traffic "not less than 3 to 5 days in advance, as a rule, notification of actions on the high seas which represent a danger to navigation or to aircraft in flight." Ballistic missile launches into the sea fall under this chapter.

1972: The United States and the Soviet Union commit in the Anti-Ballistic Missile (ABM) Treaty not to test sea-based, air-based, space-based or mobile land-based ABM systems or components. The two sides further agree not to test ABM launchers capable of firing more than one ABM interceptor missile at a time. National technical means will be used to verify these prohibitions, with each side pledging not to impede such verification.

1975: At the Conference for a Nuclear-Free Pacific, an initiative is undertaken to ban ballistic missile testing in the South Pacific.[92]

1977: As part of his "Comprehensive Proposal" to the Soviet Union, President Carter suggests limiting each the United States and Soviet Union to six ICBM and SLBM tests per year. The Soviet Union rejects the proposal.

1979: The SALT II treaty (signed by the Soviet Union and United States) encompasses a ban on: flight testing or deployment of new types of ICBMs (beyond those currently deployed), with an exception for one new type of light ICBM on each side; testing new MIRVs on existing missiles; encrypting telemetric information from test flights; production, testing and deployment of the Soviet SS-16 (because of its similarity to the IRBM SS-20 and the complications for verification that this deployment would entail); testing ICBMs from mobile launchers; flight testing air-to-surface ballistic missiles. It also calls for advance notice of ICBM launches, except for single launches not extending beyond national territory.[93] The United States had proposed a ban on testing missiles in a depressed trajectory, but dropped the proposal when the Soviet negotiators countered with proposals to limit short time of flight systems generally.[94]

1982: The Nuclear Freeze movement calls for a complete FTB in conjunction with other freeze provisions.

1986: Vanuatu, the Solomon Islands, Papua New Guinea, and Nauru seek to include a ban on missile flight testing in the South Pacific Nuclear-Free Zone Treaty. They are thwarted by Australia.[95]

1987: "Choices for Change: Security Through Arms Control," the report of a group of Democratic Congressmen, calls for a comprehensive US-Soviet FTB, with an exception to permit completion of Midgetman testing. Les Aspin, Chairman of the House Armed Services Committee, supports the measure.

1987-1988: In the Presidential campaign, five of six Democratic aspirants support a bilateral FTB.

1987: Agreement on the Establishment of Risk Reduction Centers establishes a Nuclear Risk Reduction Center in both Moscow and Washington, DC, which will be used to transmit notices called for under the 1971 and 1972 agreements listed above.

1987: The Intermediate-Range Nuclear Forces (INF) Treaty prohibits the Soviet Union and United States from testing ballistic missiles with ranges from 500-5,500 km.

1988: The Notice of ICBM and SLBM Launches Agreement provides that not less than 24 hours prior to an ICBM or SLBM launch, the Soviet Union and United States will each notify the other through the Risk Reduction Centers of the planned date of launch, missile launch area and area of impact.

1991: The START treaty prohibits encryption of telemetry (with the exception of tests related to the Strategic Defense Initiative). It calls for the broadcast of all telemetric information from ICBM and SLBM flight tests, and exchange of telemetry tapes, interpretative data and acceleration profiles for all tests. The treaty limits Russia and the United States each to 25 test silo launchers and 20 test mobile launchers at testing ranges.[96] Russia and the United States have since installed telemetry playback equipment on each other's territory.[97]

1991: In his May 29 Middle East Arms Control proposal, President Bush calls for a halt to further acquisition, production, and testing of ballistic missiles of any range by states in the region, leading to "the ultimate elimination of such missiles from their arsenals."[98]

1992: The START II treaty bans flight testing of MIRVed ICBMs after January 1, 2003.

Appendix B: Ballistic Missile Flight Test Ranges/
Space Launch Test Facilities

BM=ballistic missile flight test facilities. SLV=space launch site or flight test facility.

Argentina: Argentine press reported in 1989 that the Condor II had recently been flight tested in Patagonia.[99]

Australia: BM---Woomera. Run by Defense Science Technology Organization, Woomera reportedly has the longest recovery range in Western world. Its instrumented range covers 200 km^2 and its full range is 800 km long.[100]

Brazil: SLV---Barreira do Inferno was the first major launch facility in Brazil; Alcantara was built to launch the VLS but also has launch pads for Sonda III and Sonda IV.[101]

China: SLV---Jiuquan, Xichang, Taiyuan.

CIS: SLV/BM---Tyuratum Cosmodrome at Baikonur, Kazakhstan; Plesetsk Cosmodrome, Russia.   ICBM tests from Tyuratum fly in a north-easterly direction toward the Kamchatka Peninsula impact zone. In the east, missiles are launched from Plesetsk, the former USSR's northernmost launch facility and now Russia's main test and launch facility.

Egypt: Heliopolis?? This was the site of Egyptian missile development in the 1960s.[102]

France: SLV (Arianespace)---Kourou, French Guiana. BM---Toulon.

India: SLV---SHAR Centre (Shriharikota); BM---Chandipur (Orissa State, east coast)

Israel: SLV/BM---Palmachim Air Force Base (south of Tel Aviv).

Japan: SLV---Kagoshima.

North Korea: BM---Nodongjagu??

Pakistan: SLV/BM---Somniami Bay??

South Africa: SLV/BM---Armiston (near Overberg, in Cape Province).

United States: BM---Vandenberg AFB (California), Cape Canaveral (Florida); White Sands Army Missile Test Range (New Mexico). SLV---Kennedy Space Center (Florida), Vandenberg AFB (California), Wallops Island (Virginia)[103].

General source: "Launch Vehicles: Operational Satellite Launcher Directory," *Flight International*, April 7-13, 1993, pp. 37-41.

Appendix C: Flight Testing of Selected Developing Country Ballistic Missiles/Space Launchers

System	Date	Launch Site	Impact Site/Other Info.
BRAZIL			
Sonda III			
	1976		First launch.
	??		Second-23rd launches.
	11/30/90	Alcantra Launch Ctr.	24th launch; took a 142 kg payload to 405 km.
Sonda IV			
	1984		First launch.
	10/87		Terminated when first and second stages failed to separate due to on-board computer failure.
	??		Third launch
	4/28/89	Barreira do Inferno	Fourth launch. First test of the "hot system" to separate the two stages of the rocket, for use in the VLS.
Satellite Launch Vehicle (VLS)			
	5/18/89	Barreira do Inferno	Successful launch of a 1/3 scale version of VLS.
INDIA			
Prithvi			
	2/25/88	SHAR Centre	Success.
	9/27/89	SHAR Centre	Bay of Bengal--success.
	2/11/91	SHAR Centre	Success.
	7/4/91	SHAR Centre	Success despite bad weather.
	8/7/91	SHAR Centre	Success.
	2/92	SHAR Centre	Broke up in flight when subjected to a high-G manoeuvre to test its structural strength.
	5/5/92	Chandipur	Success.
	8/18/92	Chandipur	Success. Mobile launcher.
	8/29/92	Chandipur	Success. Clear weather. Mobile launcher.
	2/7/93	Chandipur	Success. Mobile launcher.
	6/12/93	Chandipur	Success. First from production batch. Extended range.
	11/30/93	Chandipur	Success. Impacted on island in Bay of Bengal, as intended.
Agni			
	4/20/89	Chandipur	Aborted due to problems with ignition system.

System	Date	Launch Site	Impact Site/Other Info.
	5/22/89	Chandipur	1,000 km into Bay of Bengal. Success.
	5/29/92	Chandipur	Bay of Bengal. Some problems with warhead guidance (possibly failed to detach from 2nd stage). Neither rocket nor 1 ton dummy nose cone was recovered.
	1/7/94	Chandipur	Aborted due to a technical problem.
SLV-3			
	8/10/79	SHAR Centre	Failure. Problems with stage 2 guidance.
	7/18/80	SHAR Centre	Success. 35 kg satellite (Rohini-1) into LEO.
	5/31/81	SHAR Centre	Failure. Improper orbit achieved, probably due to stage 3/4 separation anomaly.
	4/17/83	SHAR Centre	Success. 40 kg satellite (Rohini-3) carrying imaging sensors into LEO.
ASLV			
	3/24/87	SHAR Centre	Failure. Aborted after Core stage engine failed to ignite after booster separation, probably due to short circuit.
	7/13/88	SHAR Centre	Failure. Premature booster burnout leading to loss of control; deviation from flight plan; rocket break up.
	5/20/92	SHAR Centre	Success. 105 kg satellite (SROSS-C) to LEO.
PSLV[1]			
	7-8/93	SHAR Centre	Different stages of PSLV have been individually tested. Flight systems are being integrated for the launch.[2]
	9/20/93	SHAR Centre	Failure. Software glitch in guidance and control processor.[3] All four stages ignited properly.[4]
GSLV[5]			
	1995	under development at the Liquid Propulsion Test Facility at Mahendragiri.	
ISRAEL Jericho II			
	mid-70s	Iran?	Reportedly tested in Iran.
	1986	Palmachim?	Possibly two test launches.
	5/87	Palmachim?	Apparent success. Tested to a range of 820 km into the Mediterranean.
	7/6/89	Overberg Test Range, Armiston, Sth Africa	Apparent success. Into the Indian Ocean.

System	Date	Launch Site	Impact Site/Other Info.
	11/90	Overberg Test Range, Armiston, Sth Africa	
Shavit SLV			
	9/19/88	Palmachim	Success. 156 kg satellite (Ofeq-1) into LEO.
	4/3/90	Palmachim	Success. 160 kg satellite (Ofeq-2) into LEO.
Arrow ATBM			
	8/90	Mediterranean	
	3/91	Mediterranean	
	9/91	Mediterranean	
	9/23/92	Mediterranean	
	2/28/93	Mediterranean	
	11/14/93	Mediterranean	
NORTH KOREA			
Scud-B			
	1984	Nodong	Possibly three test launches
	1985	??	Possible test launch.
	1987	North of Wonsan	Possible test launch.
Scud-C			
	6/90	Nodong	Sea of Japan.
	1991	TEL in Kangwon	Sea of Japan.
	5/29-30/93	Taepo-Tong	100 km into Sea of Japan.
	5/29-3093	Taepo-Tong	100 km into Sea of Japan.
Nodong I			
	5/90	To-kol	Exploded on launch pad.
	10/91	??	Possible test launch.
	5/29-30/93	Taepo-Tong	400-500 km into Sea of Japan.

Acronyms: LEO=Low Earth Orbit; VLS=Vertical Launch System

[1] To place a 1,000 kg satellite into 900 km polar sun synchronous orbit

[2] "Two Space Launches Planned for Current Year," All Delhi Radio Network, 9 May 1993, as translated in FBIS-NES-93-088, p. 58; "ISRO Chairman on Details of Future Launches," *India Express*, 19 April 1993, as translated in FBIS-NES-93-086, p. 38.

[3] *Aerospace Daily*, 4 January 1994, p. 9.

[4] *Flight International*, 12-18 January 1994, p. 19.

[5] To place up to 2.5 ton satellite into Geostationary Transfer Orbit

Notes

1. See O. Wilkes, M. Van Frank, and P. Hayes, *Chasing Gravity's Rainbow: Kwajalein and US Ballistic Missile Testing* (Canberra Papers on Strategy and Defence No. 81, Canberra: Australia National University, 1991) Chapter 1 for a detailed description of the US Western Test Range facilities. Cost of replacing Kwajalein from *Chasing Gravity's Rainbow,* p. 15.

2. "Kwajalein Atoll Chief Sees Bright Future For Missile Facility," *Defense Daily,* 8 April 1993, p. 46.

3. See Wilkes *et al., Chasing Gravity's Rainbow,* appendix 3.

4. E.H. Kolcum, "Navy Improving Test Facilities For Trident II Missile Program," *Aviation Week & Space Technology,* 30 September 1985, pp. 76-79.

5. For a summary of technical hurdles to building a long-range missile, see L. Gronlund and D.C. Wright, "Building an ICBM," *Bulletin of the Atomic Scientists,* March 1992, p. 36.

6. F. Hussain, "The Future of Arms Control: Part IV---The Impact of Weapons Test Restrictions," *Adelphi Papers,* no. 165, London: IISS, 1981, p. 19.

7. Martin Marrietta Corporation, Aerospace Division, Special Projects Technical Report (no. TR-E-80-008), "Analysis of the Effects of Flight Test Limitations," Phase B -- Final Report, December 1980, p. 51; Wilkes *et al., Chasing Gravity's Rainbow,* p. 76.

8. Although elaborate flight testing programs along the American model are uncommon, *flights* do occur. For example, Iran and Iraq, through their extensive use of counter-city missile warfare during the late 1980s, undoubtedly gained some operational reliability and accuracy data. Ballistic missiles were also fired in the 1973 Yom Kippur war by Egypt, and over 1,000 Scud missiles were reportedly fired by the government of Afghanistan against the Mujahideen rebels.
8. The Navy is requesting 138 Trident II D5 missiles for reliability testing during the systems' projected lifetime of 30 years. This would average out to 4.5 tests per year. See prepared testimony of Rear Adm. J.T. Mitchell before the Subcommittee on Nuclear Deterrence, Arms Control and Defense Intelligence, Senate Armed Services Committee, 11 May 1993.

9. The Navy is requesting 138 Trident II D5 missiles for reliabilty testing during the system's projected lifetime of 30 years. This would average out to 4.5 tests per year. See, for example, W.S. Carus and J.S. Bermudez, Jr., "Iraq's *Al-Husayn* Missile Programme," *Jane's Soviet Intelligence Review,* June 1990, pp. 242-248.

10. See "Short-Range Ballistic Missile Infrastructure Requirements for Third World Countries," prepared by the Arnold Engineering Development Center, Arnold Air Force Base, Tennessee, Air Force Systems Command, United States Air Force (document no. AEDC-1040S-04-91), September 1991, for a thorough description of the testing infrastructure required by a developing country.

11. Afghanistan, Bulgaria, Czechoslovakia, Egypt, Hungary, Iran, Iraq, Libya, North Korea, Poland, Romania, Syria, Yemen. The Scud is a single stage, liquid-fueled rocket with a range of 300 km and a payload of 1,000 kg. It has a circular error probable of 900-1,000 meters.

12. It is less clear whether a proven design such as the Scud could be reverse-engineered or even produced indigenously from blueprints and deployed without testing. Given the potential for faulty or mismatched materials, undetected production flaws, subtleties of design not detected by the reverse engineers, and so on, it would require a leap of faith to have any confidence in the product of such an undertaking. A proliferating country would likely want to carry out at least minimal development testing to identify problems and minimal operational testing to establish functionality, even if it were receiving direct technical support from outside. See S. Flank, "Flight Test Restrictions & Reliability Analysis for Ballistic Missiles," manuscript submitted to *Science & Global Security*, for an analytical framework that could be applied to understanding the effects of flight test limitations on developing country ballistic missile programs, including modifying imported missiles.

13. For example, Janne Nolan has asserted that "an agreement to stop missile tests would slow the pace of accuracy and other technical improvements if it were negotiable, [but] its contribution to the containment of missile proliferation would be marginal." And Henry Sokolski has written, "The objective for any non-proliferation control is not detection, but rather `timely warning' of a diversion of material or technology from `safe' to `dangerous' activities (i.e., at least a year or more of warning). In this regard, missile test bans are hardly useful. Thus, a test ban would have provided little or no warning of the Iraqi development of its Al-Husayn missile, which was tested only once and was not detected before being used in war." J. Nolan, "U.S. Options for Countering the Proliferation of Ballistic Missiles: An Assessment of Possible Arms Control Measures," p. 45, in *Missile Proliferation: A Discussion of U.S. Objectives and Policy Options, CRS Report for Congress*, by R. Shuey, 21 February 1990; H. Sokolski, "Space Launch Vehicle Controls: Arguments and Answers," Aletheia Enterprises (Arlington, VA), November 1992.

14. The Prithvi is a liquid fuelled, single stage missile with a range of up to 150 km with 1,000 kg warhead, or up to 250 km with 500 kg warhead.

15. After India's tenth test of the missile, a Delhi newspaper editorial addressed the need for adequate flight testing of the system:

After ten test launches it can safely be said that India is on its way to field an indigenous SSM. The number of launches, however, is no guarantee that the system is ready for deployment in service. After the scientists are satisfied with the results, it will then be the turn of the soldiers to try the missile. Any weapons system, after all, must be thoroughly tested before those in uniform feel confident about it. It may be a tank, an aircraft or a missile, the service must ultimately give the signal that it works. Therefore, if more testing is required, so be it. A tried and tested system is preferable to one inducted in haste. This has been amply borne out by numerous instances.

The editorial called for realistic testing in adverse weather conditions: "To fully test the missile, it should be put through other launching conditions as well. That also will establish the efficacy of the guidance system." From "Editorial Hails Successful Prithvi Missile Test," *Indian Express*, 9 February 1993, as translated in FBIS-NES-93-031, p. 59.

16. On 5 July 1989, the Armaments Corporation of South Africa (ARMSCOR) announced the successful launch of a booster rocket from its Overberg test range in Cape Province. Several months later, in October 1989, NBC News reported that ARMSCOR's missile flew 1,500 km southeast into the Indian Ocean. The following day, the *Washington Post* said that "knowledgeable US officials" confirmed the NBC report and said the missile "was constructed and flown by South Africa July 5 using technology acquired from Israel."

Israeli engineers reportedly designed and built the test range. The *New York Times* added that "American satellites picked up what intelligence officials considered to be an important piece of information: the rocket plume of the South African missile bore a striking resemblance to that of Israel's Jericho missile.... Administration officials also say that equipment seen at the South African missile test resembles equipment used by the Israelis in their own missile tests." From R.J. Smith, "Israel Said to Help S. Africa on Missile," *Washington Post*, 26 October 1989, p. A36; B. Gertz, "S. Africa on the Brink of Ballistic Missile Test," *Washington Times,* 20 June 1989, p. 1; *Narodna Armiya*, 15 September 1989, as translated in FBIS-EEU 21 September 1989; Johannesburg domestic service, 5 July 1989, as translated in JPRS-TAC 19 July 1989; M.R. Gordon, "U.S. Says Data Suggest Israel Aids South Africa on Missile," *New York Times,* 27 October 1989, p. A1; "State Confirms Discussions with Israel on Pretoria Cooperation," *Aerospace Daily*, 27 October 1989, p. 155.

17. D.A. Fulghum and J.M. Lenorovitz, "Israeli Missile Base Hidden in Hill," *Aviation Week & Space Technology*, 8 November 1993, p. 29.

18. James. N. Constant writes in *Fundamentals of Strategic Weapons: Offense and Defense Systems* (Martinus Nijhoff Publishers, 1981): "The accuracy of a missile is probably the most sensitive element of its effectiveness... At best however, the accuracy of a missile can be defined only as a probability since many error sources of both a systematic and random nature can combine to produce a miss by the missile at the target. Consequently, a very large number of missiles must be tested before a given level of confidence can be obtained that a given missile, or a given class or type of missile, will indeed hit its target within some reasonable distance to ensure target destruction. Improving the missile accuracy therefore requires a large investment in missile test ranges, apart from the cost of missiles used in the tests." (p. 167)

19. A. Karp, "Ballistic Missile Proliferation," *SIPRI Yearbook 1991: World Armaments and Disarmament* (Stockholm: SIPRI, 1991), p. 319; J.S. Bermudez, Jr., "New Developments in North Korean Missile Program," *Jane's Soviet Intelligence Review*, August 1990, p. 343.

20. Prepared testimony of R.J. Woolsey before the Senate Government Affairs Committee, 24 February 1993.

21. See D.C. Wright and T. Kadyshev, "An Analysis of the North Korean Nodong Missile," *Science & Global Security*, Vol. 4 (forthcoming), pp. 129*ff.*

22. See, for example, KYODO, 11 June 1993, as in FBIS-EASS-93-111, p. 4.

23. AFP in *Chicago Sun-Times*, 20 April 1993, p. 46.

24. US Defense Support Program (DSP) satellites equipped with infrared sensors reportedly detected all Iraqi Scud launches during the 1991 Gulf War. In addition, Airborne Warning and Control System (AWACS) aircraft, tactical reconnaissance aircraft and the E-8A Joint Surveillance Target Attack Radar System (JSTARS) detected missile activity during the war.

Several sensors are currently under development in the United States to detect third world missiles. For example, Los Alamos National Lab is developing a transportable light detection and ranging (LIDAR) system which can rapidly and accurately identify missile exhaust plumes. In addition, the US Air Force is contracting out for the creation of a data base on radar measurements of the exhaust plume of various missiles. See US Department of Defense, *Report to Congress on the Conduct of the Persian Gulf War*, September 1991; C. Covault, "Reconnaissance Satellites Lead Allied Intelligence Effort," *Aviation Week & Space Technology*, 4 February 1991; R.J. Smith, "Compactness, Simplicity of Iraq's Scuds Complicate US Search," *Washington Post*, 20 January 1991; R.A. Mason, "The Air War in the Gulf," *Survival*, No. 33 (May/June 1991); J.

Boatman and B. Starr, "Eyes of the Storm," *Jane's Defense Weekly*, 4 May 1991; D.A. Fulghum, "Lasers Track Missile Plumes," *Aviation Week & Space Technology*, 24 January 1994, p. 40; "USAF Eyes Contract Award for Missile Pluume Data Base," *Inside the Air Force*, 23 July 1993, p. 11.

25. Hussain, "The Future of Arms Control," p. 31 and footnote 24.

26. Ibid., footnotes 36-40.

27. R. Sherman, "Deterrence Through a Ballistic Missile Flight Test Ban," *Arms Control Today*, December 1987, p. 8.

28. A. Bolonkin, *The Development of Soviet Rocket Engines (For Strategic Missiles)*, Delphic Associates Inc. (Falls Church, VA), 1991, pp. 100-108.

29. MTCR members as of June 1993 are: Argentina, Australia, Austria, Belgium, Canada, Denmark, Finland, France, Germany, Greece, Hungary, Iceland, Ireland, Italy, Japan, Luxembourg, the Netherlands, New Zealand, Norway, Portugal, Spain, Sweden, Switzerland, the United Kingdom, and the United States. South Africa is slated to become a member at the next MTCR meeting in October 1994. Russia and Israel have agreed to adhere to the export guidelines but are not formal members of the regime.

30. "Missile Technology Control Regime: Fact Sheet to Accompany Public Announcement," White House Office of the Press Secretary, 16 April 1987, p. 1.

31. Nolan, *Missile Proliferation*.

32. The presence of hundreds of ballistic missiles in the Middle East has not deterred war or the use of these missiles. In fact, only Afghanistan and the Middle East have seen ballistic missile warfare since Germany fired V-2 missiles in World War II. In the 1973 October War, the intense missile warfare during the Iran-Iraq war, the Afghan civil war, and the war over Kuwait, between two and three thousand Scud missiles were fired. This pattern of missile use is especially ominous when considering the possible proliferation of mass destruction payloads in the region. Unconventionally armed missiles might be considered useable, rather than as deterrents only.

33. R.J. Smith, "U.S. Urged to Cut 50% of A-Arms," *Washington Post*, 6 January 1992, p. 1.

34. "DOD Eyes Mini-Nukes, Ballistic Missiles for Counterproliferation," *Inside the Pentagon*, 16 December 1993, pp. 1, 8-10.

35. R. Holzer, "US Navy Targets Conventional Deterrence," *Defense News*, 10-16 May 1993, p. 6

36. R. Holzer and G. Leopold, "US Navy Girds for New Threats," *Defense News*, 8-14 March 1993, pp. 1, 28.

37. "Navy Offers Submarines for Conventional Deterrence," *Defense Daily*, 21 May 1993, p. 295.

38. R. Holzer and N. Munro, "US Navy Tests Non-Nuclear Trident," *Defense News*, 13-19 December 1993, p. 4; J. Moag, "Navy to Test GPS-Guided D-5 Trident Missile Off SSBN Nebraska," *Inside the Pentagon*, 18 November 1993, pp. 1, 6.

39. See Appendix A for summary of negotiated and proposed flight test restrictions.

40. Nolan, *Missile Proliferation*.

41. A. Ignatius, "Russian Lawmakers Opposed to Start II Pledge Fight as Hearings on Pact Begin," Wall Street Journal, 3 March 1993, p. A11.

42. "Nuclear Notebook," *Bulletin of the Atomic Scientists*, March 1993, p. 49.

43. "Russian Nuclear Modernization Plans Could Jeopardize Nunn-Lugar Aid," *Inside the Pentagon*, 14 January 1993, pp. 2-3; "CIA expects Russia to deploy three new ballistic missiles by 2000," *Aerospace Daily*, 4 February 1993, p. 195.

44. "New ICBM Tests Set for 1994," *Kuranty*, 8 April 1993, as translated in FBIS-SOV-93-067, p. 62.

45. Prepared testimony of R.J. Woolsey before the Senate Foreign Relations Committee, 24 June 1993.

46. "Experimental `START' Rocket Launched," ITAR-TASS, 25 March 1993, as translated in FBIS-SOV-93-057.

47. J.M. Lenorovitz, "US Entrepreneurs Seek Russian SLBMs," *Aviation Week & Space Technology*, 19 April 1993, p. 22-23; "Missiles may become satellite launchers," *Jane's Defence Weekly*, 15 May 1993, p. 11.

48. See P.S. Clark, "Converting Soviet Missiles into Russian Space Launchers," *Jane's Intelligence Review*, September 1993, pp. 401-404.

49. A facile presumption that the declared end of the Cold War has aborted any such race would need to be supported by the actual cancellation of all such R&D projects and dispersal of the infrastructure for carrying them out. Such an assumption also begs the question of whether the arms race was solely a product of ideological conflict.

50. Depressed trajectory (DT) equals the minimum flight time for a given distance. This trajectory could be used to achieve a first strike or to penetrate missile defenses. DT would necessitate new hardware--if not a new missile at least a new RV, which would have to be tested. L. Gronlund and D.C. Wright,

"Depressed Trajectory SLBMs...," *Science and Global Security*, Vol. 3 No. 1-2 (1992).

51. Again, this issue might be dismissed by some as irrelevant to the post-Cold War world. But as long as the great powers retain, and continue to modernize, nuclear warfighting capabilities, we can only guess at how many weeks of "serious international crisis" it would take to get statesmen started again on "thinking the unthinkable."

52. Sherman, *Arms Control Today*, p. 9.

53. Slocombe also argued against a total ballistic missile flight test ban because it "would block important programs needed to maintain stable deterrence, would present a number of verification problems, and would divert the superpowers from pursuing more fruitful and effective arms control measures." See W.B. Slocombe, "A Flighty Idea," *Arms Control Today*, December 1987, pp. 14-15.

54. Chart with the prepared testimony of Rear Adm. J.T. Mitchell before the Subcommittee on Nuclear Deterrence, Arms Control and Defense Intelligence, Senate Armed Services Committee, 11 May 1993.

55. See Flank, "Flight Test Restrictions and Reliability Analysis for Ballistic Missiles."

56. Hussain, "The Future of Arms Control," p. 28.

57. Hussain may have been referring specifically to the hardening of RVs against nuclear effects and the development of other countermeasures to Soviet ABM systems, an active area of US research in the late 1960s and early 1970s. Likewise the development of effective countermeasures to new ballistic missile defense systems would probably depend on flight testing, though perhaps not as strongly as the development of complete new missile systems. However, the hardening of silos is achieved through passive measures and the hiding of submarines is also ensured through measures that do not involve missile flight testing. A ballistic missile sitting in a silo or rising through the atmosphere is inherently a soft target and cannot be substantially hardened without adding a prohibitive burden of dead weight.

58. See, for example, Martin Marietta Corporation, Aerospace Division, Special Projects Technical Report (no. TR-E-80-008), "Analysis of the Effects of Flight Test Limitations," Phase B -- Final Report, December 1980, p. 57, released under US Freedom of Information Act to Nautilus Institute.

59. Wilkes *et al.*, *Chasing Gravity's Rainbow*, pp. 9-13.

60. Private communication, from M.C. Cleary, Chief, History Office, the Department of the Air Force, Headquarters 45th Space Wing (AFSPACECOM), Patrick Air Force Base, Florida, 31 January 1994.

61. Chart with the prepared testimony of Rear Adm. J.T. Mitchell before the Subcommittee on Nuclear Deterrence, Arms Control and Defense Intelligence, Senate Armed Services Committee, 11 May 1993.

62. Ibid..

63. Ibid..

64. US General Accounting Office, *ICBM Modernization: Minuteman III Guidance Replacement Program Has Not Been Adequately Justified*, report no. GAO/NSIAD-93-181, June 1993; E. Grossman, "Butler: Delay in ICBM Propulsion Replacement Adds `Significant Risk'," *Inside the Pentagon*, 22 October 1993, pp. 1, 10.

65. The M-4 is a three-stage SLBM with a range of up to 4,000 km. It underwent 14 developmental flight tests, with only one failure. The last developmental test was completed on 29 February 1984, and the missile entered into qualification/acceptance phase shortly thereafter. *Aviation Week & Space Technology*, 22 June 1981, p. 23; *Aviation Week & Space Technology*, 9 April 1984, p. 65.

66. *Le Monde*, 17 February 1993, as translated in JPRS-TAC-93-006, p. 15.

67. Ibid.

68. Wilkes *et al.*, *Chasing Gravity's Rainbow*, p. 128, footnote 14.

69. "Britain's Trident Cuts Won't Affect Lockheed's D-5 Totals, Rifkind Says," *Aerospace Daily*, 18 November 1993, p. 289; "Britain Reduces Trident Firepower," *Defense Daily*, 18 November 1993, p. 257.

70. J.W. Lewis and Hua Di, "China's Ballistic Missile Programs: Technologies, Strategies, Goals," *International Security*, Fall 1992 (Vol. 17, No. 2), p. 29.

71. Ibid., p. 30.

72. Prepared testimony of William Webster before the Senate Governmental Affairs Committee, 18 May 1989.

73. Although little discussed, restrictions on static testing of rockets could also be used as a verifiable measure of compliance with a pledge to forgo missile development. Static tests of rocket motors, which generate vast clouds of hot exhaust gases, are usually conducted in the open. Both infrared and chemical signatures, as well as direct observation of test stand facilities, would make possible their detection. Hiding or camouflaging such tests would not be

impossible, but would require costly facilities that would be themselves vulnerable to detection by reconnaissance or human intelligence. Implementation of such restrictions in combination with a flight test ban could provide a useful confidence-building measure. See J.M. Lenorovitz, "MSTI-1 Satellite Images Rocket Motor Test Fire," *Aviation Week & Space Technology*, 22 February 1993, p. 64.

74. Slocombe, for example, writes: "A ballistic missile flight test ban faces serious verification problems. A considerable amount of information relevant to improving existing ballistic missiles could be derived from tests disguised as space tests of various kinds...[A]n extended development program using disguised tests could be followed by a deliberate breakout, doing a quick series of openly violative tests to complete the work. Requiring that space launchers be distinct from military missiles would entail very large costs, and even so would not completely solve the problem, because of inevitable overlaps of the technology." Slocombe, *Arms Control Today*.

75. Sherman, *Arms Control Today*, p. 9

76. See Ibid., pp. 9-10; P. Zimmerman, chapter 11 of this text; J. Scheffran, "Verification of Missile Bans/Monitoring of Space Launches," prepared for INESAP conference, Mühlheim, Germany, 27-31 August 1993.

77. Wilkes *et al.*, *Chasing Gravity's Rainbow*, p. xiv.

78. In addition, it may be possible to verify through national technical means the *non-production* of solid-fueled rockets. See V. Thomas, "Monitoring Solid-Fueled Missile Production for Arms Control," *Physics and Society*, Vol. 17 No. 1 (January 1988), pp. 8-10.

79. S.D. Drell and T.J. Ralston, "Restrictions on Weapons Tests as Confidence-Building Measures," in B. Blechman, ed. *Preventing Nuclear War: A Realistic Approach*, Bloomington, IN: Indiana University Press, 1985.

80. Hussain, "The Future of Arms Control," pp. 29, 33.

81. *Jane's Defence Weekly* reported in its 1 May 1993 issue that Russia has revealed a new version of its infamous Scud missile with "a highly accurate guided warhead." The missile, known as the Scud-B Mod 2, was developed in the late 1980s. Carrying a conventional submunition warhead, it is believed to have entered into service with the Russian Army.

82. K. Adelman, "Curing Missile Measles," *Washington Times*, 17 April 1989, p. D1 and "How to Limit Everybody's Missiles," *New York Times*, 7 April 1991; K.C. Bailey, "Rushing to Build Missiles," *Washington Post*, 6 April 1990, p. A15 and "Can Missile Proliferation Be Reversed?," *Orbis*, vol. 35 no. 1 (Winter 1991), pp. 5-14.

83. The continental United States is roughly 8,000-10,000 km from the Middle East and 8,000 km from the Korean peninsula.

84. Wilkes *et al.*, *Chasing Gravity's Rainbow*, pp. 122-123.

85. Sixteen countries deploy or have until recently deployed short range ballistic missiles (110-500 km): Afghanistan, Pakistan, North Korea, South Korea, Egypt, India, Iran, Libya, Syria, Yemen, Germany, Italy, the Netherlands, Hungary, Poland and Romania. Of these sixteen, six are European countries that have already or will shortly demobilize their short-range Lance, Scud, SS-21 and SS-23 missiles. Of the remaining ten---all of which are developing countries--four deploy only Scud missiles transferred from the Soviet Union in the 1970s-80s and have no known indigenous missile production capability or intent. Five of the remaining six imported Scud or other short-range missiles and are now producing them or attempting to do so. Only India has indigenously developed its short-range Prithvi missile.

Four additional countries deploy or have until recently deployed intermediate range (500-5,500 km) ballistic missiles: Bulgaria (500 km), Czechoslovakia (500 km), Israel (1,450 km), and Saudi Arabia (1,850 km). North Korea is working on a 600-1,000 km range missile, with only one successful flight test reported. India has flight tested a 2,400 km range missile, the Agni, three times. Iraq previously deployed 600 km and possibly 800 km extended-range Scud missiles. These missiles have been destroyed by the UN Special Commission implementing the Gulf War cease-fire agreement.

Currently eight countries deploy intercontinental-range ballistic missiles: Belarus, China, France, Kazakhstan, Russia, Ukraine, United Kingdom and United States.

86. "US Presses Mideast Missile Talks," *Washington Post*, 28 December 1988, p. 15.

87. White House, Office of the Press Secretary, "Fact Sheet on Middle East Arms Control Initiative," 29 May 1991, p. 2.

88. MENA (Cairo), 5 August 1991, as translated in FBIS-NES-91-151, p. 20.

89. Prepared testimony of R.J. Woolsey before the Senate Foreign Relations Committee, 24 June 1993, p. 4.

90. See Gronlund and Wright, *Science and Global Security*.

91. Hussain, *Adelphi Papers*, footnote 43.

92. Wilkes *et al.*, *Chasing Gravity's Rainbow*, p. 122.

93. US Arms Control and Disarmament Agency, *Arms Control and Disarmament Agreements: Texts and Histories of the Negotiations*, Washington, DC: USGPO, 1990.

94. See Gronlund and Wright, *Science & Global Security*, pp. 133-135 for more on the history of proposals to limit or ban DT flight testing.

95. Wilkes *et al.*, *Chasing Gravity's Rainbow*, pp. 123-128.

96. Wilkes *et al.*, *Chasing Gravity's Rainbow*, p. 4 citing US ACDA, "Strategic Arms Reduction Talks," Issue Brief, 29 July 1991.

97. Prepared testimony of R.J. Woolsey before the Senate Foreign Relations Committee, 24 June 1993.

98. White House, Office of the Press Secretary, "Fact Sheet on Middle East Arms Control Initiative," 29 May 1991, p. 2.

99. *Noticias Argentinas*, 4 April 1989, as translated in FBIS-LAT 5 April 1989, p. 32.

100. M. Richardson, "Need a Missile Testing Range?," *International Herald Tribune*, 7-8 April 1990.

101. M.J. Collins, *Evaluating the Military Potential of a Developing Nation's Space Program: A case study of Brazil*, September 1991, Naval Postgraduate School, Monterey, CA.

102. T. Prittie, "Bomb Shop in the Nile: Target Israel," *Atlantic Monthly*, Vol. 214 No. 2 (1964), p. 38.

103. In 1986 General Dynamics obtained permission from NASA and DOT to launch from Wallops Island. A launch occurred in May 1993.

10

Land-Based Ballistic
Missile Verification:
The UNSCOM Experience

Timothy V. McCarthy[1]

During the short history of the United Nations Special Commission on Iraq (UNSCOM), UN inspectors have destroyed, removed or rendered harmless the vast majority of Baghdad's nuclear, chemical, and biological weapon and missile delivery capabilities, as well as a variety of research and production facilities and related equipment. UNSCOM's mandate, however, will not end once these capabilities have fully been eliminated, for the Security Council has instructed UNSCOM to ensure that Baghdad's weapons of mass destruction (WMD) programs are never reconfigured. To implement the Council's directive, UNSCOM has developed a plan for the Ongoing Monitoring and Verification (OMV) of Iraq which will rely heavily on intrusive On-Site Inspection (OSI) technique--as have UNSCOM's inspections to date. The Commission's inspection efforts and its OMV plan break significant new ground in global efforts to control the spread of unconventional weaponry.

Based on successes enjoyed in the disarmament of Iraq, it has become fashionable to assert that UNSCOM's OSI role should be expanded or extended beyond the Iraqi case. So far, few attempts have been made to assess UNSCOM from an organizational perspective[2]--that is, to examine the Commission's legal basis, structure, budget, intelligence functions, and other such issues--to more carefully determine its potential utility in monitoring and verifying future arms control and non-proliferation agreements outside of the Iraqi case.[3] Moreover, the OMV plan itself has received scant attention in the literature, which is surprising given that it provides a convenient model upon which other multilateral verification techniques and modalities may be based. In this study, I examine UNSCOM as an organization and the OMV plan as a model in order to reach some conclusions about the role that UNSCOM or a similar UN body can play in missile nonproliferation efforts. Although I focus on terrestrial monitoring techniques (such as OSI), many of the study's findings may also be applied to non-terrestrial monitoring instruments (for example, satellite monitoring).[4]

I proceed as follows. First I identify a specific issue (such as personnel selection or the OMV plan) and describe UNSCOM's experience with, and attempts to solve problems related to, that issue. Next, I discuss the general implications of that experience for further UN monitoring and verification efforts, which is followed by more specific application to one proposed missile nonproliferation regime, the Zero Ballistic Missile (ZBM) proposals put forth by the Federation of American Scientists (see chapter 9). I conclude the study by deriving several important lessons from the UNSCOM experience and indicate several areas where additional research is needed.

The ZBM proposal involves two basic points: a ban on all missiles with greater than a 100km range when carrying a 500kg warhead, and the allowance for peaceful Space Launch Vehicle (SLV) development.[5] I add one further explicit assumption, that is, that all states party to the agreement have a right to participate in the monitoring, verification and compliance of the agreement.[6] My purpose in applying UNSCOM's organizational and other experiences to the ZBM proposal is not to determine the plausibility or utility of a global ban, but rather to explore the difficult and complex problems that might arise with any multilateral monitoring and verification exercise.[7]

One major analytical difficulty must be highlighted from the outset: due to Baghdad's unwillingness to accept provisions of certain Security Council resolutions, formal OMV efforts in Iraq have yet to begin. In the meantime, UNSCOM implemented what it terms the "Interim Monitoring and Verification" (IMV) of Iraqi missile programs and related facilities which, along with the inspections that have already taken place and the outline of the OMV plan itself, provide significant data to make preliminary assessments of UNSCOM's monitoring and verification experience. A more complete analysis awaits the actual and continued implementation of the OMV regime.

Resolution 687

Any discussion of UNSCOM begins with *Resolution 687*, adopted by the Security Council on April 3, 1991, a little over one month after fighting in the Gulf had ended. Originally drafted by the US Government (primarily the State Department)[8] and enacted under the enforcement provisions of the UN charter, *687* is best described as a conditional ceasefire which offered a cessation of hostilities in exchange for Iraqi acceptance of clearly defined disarmament and other conditions. As such, *687* does not represent a contractual arrangement between parties, but rather an imposition of directives upon a defeated power. In other words, Iraq has no legal basis to unilaterally abrogate its *687* obligations.[9]

Several of the resolution's provisions are noteworthy. First, the Council demanded that Iraq "unconditionally accept destruction, removal or rendering harmless...of all ballistic missiles with a range greater than 150km and related major parts, and repair and production facilities," and to "undertake not to use, develop, construct or acquire" such items in the future. Similar directives apply to Baghdad's nuclear, biological and chemical (NBC) weapon programs.[10] To implement the various disarmament provisions, *687* created a Special Commission, acting under the Council's authority, to conduct immediate on-site inspection of Iraqi missile capabilities "based on Iraqi declarations and the designation of any additional locations by the Special Commission itself."

Second, the Council called for member states to continue to prevent sales of arms, or any technologies or systems related to NBC or missile capability to Iraq, though it agreed to review this and other embargo provisions (excepting prohibitions against NBC technology or missile sales) on a regular basis, "taking into account Iraq's compliance with the resolution and general progress towards the control of armaments in the region."

Third, *687* notes that once the Council agrees that Iraq has destroyed, removed or rendered harmless its unconventional weapon capabilities and agreed not to acquire them in the future, the prohibition against the "import of commodities and products originating in Iraq [most importantly, oil]...shall have no further force or effect." And fourth, the Council notes that Iraqi unconventional disarmament would represent a step "towards the goal of establishing in the Middle East a zone free of weapons of mass destruction and all missiles for their delivery."

Significance

To examine the significance of *687* for missile nonproliferation, it is useful to address two questions. Does *687* establish a legal precedent for extension of the UN's role in nonproliferation? And, what is the nature of the compliance strategies embodied in *687*?

Turning the first question on its head, there is little evidence that the Council felt it could draw on *prior* Council directives as precedent for the specific resolution itself. During debates on *687*'s adoption, no delegation made any specific reference to previous Security Council action with respect to disarmament; indeed, several members referred to the resolution as "unique" or "unprecedented."[11]

As to future precedent, *687* may have a more *de facto* than *de jure* relevance. Of course, in the event that another unconventionally armed aggressor--who has demonstrated a willingness to use WMD--is defeated pursuant to Security Council authorization, a similar resolution could very easily be adopted. Clearly the reference point then would be *687.* Secretary General Boutros Boutros-Ghali has recognized that what might be termed "enforced disarmament" will become one of the UN's future responsibilities.[12] Some of *687*'s language may prove useful as well, especially the "destruction, removal, rendering harmless" and the "undertake not to use, develop, construct or acquire" clauses.

However, given the unique nature of the Gulf conflict, similar cases of coercive destruction of missile forces will be few and far between.[13] It will be exceedingly difficult therefore to apply legal rationales found in *687* to negotiated missile reductions between equal partner states. As Tim Trevan, special advisor to UNSCOM chairman Rolf Ekeus, has noted, the UNSCOM experience derives from a different "point of departure...it is an enforcement measure, not an international agreement."[14]

One should not discount *687*'s relevance out of hand, however, for indeed the resolution provides ample *de facto* precedent for a continued UN role in nonproliferation efforts. Perhaps most important, the simple fact that *687* created a UN inspectorate actively involved in missile nonproliferation efforts changes the terms of debate regarding possibilities for such efforts. UNSCOM has pulled multilateral verification out of theoretical textbooks and implemented it in practice. Now, there is a concrete model to study and debate. Moreover, the designation of UNSCOM as a subsidiary body of the Council closely ties UN monitoring and verification responsibilities to the body whose primary task is to deal with threats to international peace and security. Indeed, *687* explicitly accepts the notion that the mere possession of WMD in the hands of an aggressor state threatens international peace and stability, and that the Council may take concerted action in such a case.[15]

Following in these footsteps, the Council's January 31, 1992 meeting--the first meeting of the Security Council at the level of Heads of State and governments--is of particular importance. The Council's official statement from that meeting voiced concern over the proliferation of weaponry and read, *inter alia*, that "proliferation of all weapons of mass destruction constitutes a threat to international peace and security" and that the members of the Council "commit themselves to working to prevent the spread of technology related to research for or production of such weapons and to take appropriate action to that end."[16] The statement must have been made with the Iraqi experience and the precedent of *687* in mind. And although the Council's statement does not have the legal force of a resolution, it is one step away from again using the Council's enforcement authority to deal with the proliferation of dangerous weaponry, including that of missiles and related technologies.

Two points should be made with respect to *687*'s compliance strategies. First, *687* embraces a fairly straightforward carrot-and-stick approach, that is, the use of negative incentives (maintenance of oil sales and commodity sales embargo and implicitly, threat or actual use of force) and positive incentives (lifting of the embargoes) to affect Iraqi compliance. Thus, the resolution provides Baghdad with an important stake in the disarmament process. Second, the resolution explicitly ties compliant or non-compliant behavior with specific Security Council responses. This latter point will be most important in any continuation of the UN's role as an active participant in arms control or nonproliferation agreements.

Finally, it is interesting to note that the Council appeared to view, at an early date, the Special Commission as not only an international inspectorate, but a body with broader interpretive and intelligence powers. In order to "designate additional locations" it is clear that UNSCOM would need to be capable of receiving information from member states on Iraqi weapon programs, generating its own data, and providing in-house analysis for additional inspections.

ZBM Application

Several of *687*'s provisions would be applicable to UN monitoring, verification, and compliance of a ZBM. As mentioned above, the specific language related to prohibited activities could well be incorporated into a ZBM treaty, though language referring to prohibition of space technology diversion to military purposes would most certainly have to be added. *687*'s demand for information exchange--that is, Iraq's obligation to provide information to the Council on all its WMD and missile facilities--would also be central to a ZBM, as it has been for other missile limitation regimes such as INF or START. In more general terms, the concept of creating a special international body (such as UNSCOM) acting under Council auspices, specifically to conduct OSI based on both signatory declarations and additional designations by the special body, could be carried over.

It remains obvious, however, that many of *687*'s provisions would not be applicable to a ZBM, primarily due to the latter's proposed status as an internationally negotiated treaty rather than an imposed cease-fire condition. Foremost among these would be *687*'s implicit use of force--since the relationship between non-compliance and the use of force will unlikely be as closely related--and the unconditional language (for example, "the Council *demands* Iraq...) contained in the resolution. Also, non-compliance penalties might include the *imposition* of an international embargo of one type or another, rather than the *maintenance* of an already existing embargo as in the Iraqi case.

There is one interesting point to be made, however, with respect to use of force provisions in a ZBM. In a treaty that bans any long-range missiles, any breakout scenario--an ICBM developed from a SLV, for example--would clearly represent a "threat to international peace and stability." Particularly in the absence of effective missile defenses, military action undertaken in defense of the treaty cannot entirely be ruled out. Of course, compliance enforcement (especially the use of force) is ultimately the responsibility of sovereign states, and any such response is far from being a certainty. The resolution of the North Korea-NPT crisis will set an important precedent in this regard.

UNSCOM's Structure and Decision-Making

Although its core elements have remained fairly constant over time, UNSCOM has been a dynamic organization since its inception. It has adapted according to a perceived need or political reality. The description that follows reflects UNSCOM as it is currently constituted. My observations with respect to the OMV plan will reflect changes to the Commission that are envisioned by its officials and analysts.

Structure

UNSCOM may be thought of as two inter-related sub-groups: the working level organization, which plans and executes inspections and reports directly to the Security Council; and the formal Special Commission itself, which does not have any day-to-day responsibilities.

Unlike the collegial, consensual bodies that tend to characterize UN-established bureaucracies, the "working level" headquarters in New York is a small executive body able to quickly make and implement decisions.[17] At its apex is the Executive Chairman, Rolf Ekeus, who is assisted in the Office of the Chairman by a Deputy Chairman, a Legal Advisor a Political Advisor and two support staff. The Office of the Chairman is supported by a small Administrative office of two professionals and six support staff.

Permanent staff of the Division of Operations consists of seven advisors with nuclear, chemical biological and missile expertise. Operations is charged with conducting and providing logistical support for OSIs. Inspections to date included hundreds of different inspectors on loan from 38 member states.

The Information Assessment Unit (IAU), also located at the New York headquarters, did not begin full operation until January 1992, nearly a year after the establishment of UNSCOM.[18] As the name implies, the Assessment Unit's primary functions are, *inter alia*, to provide analysis in support of inspections, to coordinate UNSCOM's information and document flow, and to examine and verify Iraqi weapon program declarations. The nine analysts within IAU are weapons experts in the nuclear, biological, chemical and missile fields, all of whom have also served either on inspection or interim monitoring teams; an American and a Frenchman comprise the ballistic missile team. The intelligence functions of the Assessment Unit are more fully discussed below.

UNSCOM also maintains 24 personnel in its Bahrain field office, including an air crew and 84 personnel in the Baghdad field office, including members of the chemical weapon destruction group and helicopter and maintenance crews.[19] Thus, UNSCOM working level personnel currently total 140 persons.[20]

The Special Commission itself is composed of 19 members (not including UNSCOM's Chairman and Deputy Chairman) who are appointed by the Secretary General. In light of the preponderance of western industrialized experts at the working level and the subsequent need to obtain a multinational bent to the Commission, commissioners represent 19 different member states, including representatives from Venezuela, Indonesia, China and Nigeria.

The Commission meets as a fully constituted body for plenary sessions twice per year; there have been five plenaries to date, the last having occurred in May of this year. The plenaries review UNSCOM activities over a certain defined time frame. Commissioners, who are generally scientific as opposed to weapons experts, sit on several advisory boards or working groups: nuclear, biological and chemical, ballistic missile, and future compliance and monitoring. A destruction advisory panel (dealing primarily with chemical weapons) assists the Commissioners and UNSCOM in execution of their duties. The missile advisory group is headed by an Italian with the other member an Indonesian military officer.

According to UNSCOM officials and analysts, the Commission serves primarily to "legitimize" UNSCOM activities by virtue of its national diversity. The Commission has no formal decision-making authority and is not an active interface between working-level headquarters and the Security Council. Essentially, the Commission "looks good" in the eyes of member states and for the "UN culture" as a whole.

In formal session, the Commission receives briefings on inspections, administration and other problems from UNSCOM personnel. As a corporate body, therefore, the Commission has had little operational influence, although as individuals, various members have had an impact; several members have gone on inspections while the destruction advisory panel conceived the chemical weapons (CW) destruction plan. The advisory boards do submit written recommendations on operational and other matters, but often they are not timely or have been implemented already by working level teams.[21]

Decision-Making

UNSCOM is an executive, rather than a deliberative or consensual body. Chairman Ekeus has all day-to-day decision-making authority and judged by UN standards, has tremendous power and influence. By nearly all accounts, the success of UNSCOM may to a large degree be traced to the executive structure and Ekeus's effective leadership and management style. The chairman maintains tight control over the workings of UNSCOM and tends to deal individually with the headquarters personnel. Most importantly, there is no intervening level of management between Ekeus and the Security Council; in other words, the Chairman does not report directly to the Secretary General. Ekeus is able to quickly address compliance issues directly with those in charge of enforcing decisions.

Significance

UNSCOM's status as a subsidiary body to the Security Council rather than the Secretary General is the single most important lesson to apply for further UN monitoring and verification tasks. Not only does a monitoring and verifying body fall logically under the Council's international security responsibilities, but avoids, by and large, being bogged down in the General Assembly. To the extent possible, long-winded deliberations on compliance or breakout issues must be avoided in any future multilateral nonproliferation regime; a closer association with the Secretary General or the General Assembly would almost assure that such deliberations took place.

Aside from the ability to quickly address compliance issues, a closer association with the Security Council rather than the General Assembly may preserve the flexibility and independence necessary to undertake complex monitoring and verification tasks. It is doubtful that UNSCOM's own "esprit de corps" could have been maintained if it was closely linked with the UN's regularized logistic and administrative support structures. Although UNSCOM does receive personnel and other support from other UN agencies (primarily from the Secretariat and the Office for Disarmament Affairs), and maintains its headquarters within the UN building, it remains largely separated from the rest of the UN bureaucracy in terms of its high profile, and nature of operations. As one inspector told the author, UNSCOM is a "maverick" organization with "permission to break rules." While breaking rules may not be the order of the day, it is clear that an international inspectorate must avoid being swept into the bowels of the UN bureaucracy.

ZBM Implications

Maintaining UNSCOM's basic organizational structure and, most importantly, its relatively small *executive* body as a base for monitoring and verification operations would prove to be an acceptable model for the ZBM to follow. Of course, given the larger number of countries likely involved in such a regime, the number of analysts at headquarters and inspectors in the field would need to be increased.

A ZBM inspectorate would need to make one significant change from the way in which UNSCOM currently operates. That is, the Commission, or some similar body, would need to be given broader authority to honor the principle that all states party to a missile treaty would maintain the right to participate in its verification. Such additional authority might include the right to conduct independent reviews of inspection activities, the raising of the Commission's profile to serve as a true interface between the inspectorate and the General Assembly; and additional responsibilities in determining inspection modalities. Membership might also be limited to those countries not represented on the inspection teams and be on a rotating, rather than permanent basis. In such an arrangement, compliance decisions would still be made by the Security Council and day-to-day operations would still be controlled at the working level. The Commission would then become less of a decision-making body than an Inspector General, for example, monitoring inspections, budgets, etc. Finally, the responsibility for determining who will actually serve on the Commission would likely be shifted from the Secretary General's office to the national decision-making authority of states party to the treaty.

Resources: Budget and Personnel

Budgets

Financing UNSCOM operations has proven to be a complex and controversial issue. UNSCOM did not even have a formal budget well into its eighth month of operation. The Secretary General did submit a budget to the General Assembly, but the proposal was rejected because it was based on financing from fixed member state assessments. As a result, UNSCOM has had to rely largely on donations from those states willing to provide cash or contributions in-kind. This lack of fixed sources of funding has caused difficulties with long-term planning and staffing and could prove particularly troublesome as the inspection phase gives way to the long-term OMV regime.[22]

To be sure, the Security Council has explicitly addressed the problem of UNSCOM financing. Under Resolution 699, adopted June 17, 1991, Iraq is fully liable for tasks undertaken by the Special Commission. Resolution 706, adopted August 15, 1991, allowed for the closely monitored sales of Iraqi oil to pay for, *inter alia*, UNSCOM's operations. For a variety of reasons, the Iraqis have refused to sell any oil for these purposes and so the Council adopted Resolution 778 on October 2, 1992 to establish an escrow account to release Iraqi frozen assets held by member states. UNSCOM has only recently begun to use these frozen assets to defray its costs.

The total of the sums involved in UNSCOM operations has been sizeable, thought not extraordinary: from the onset of operations in May 1991 through December 1992, costs totalled US$26.4 million, $17 million of which is for travel-related expenses. An estimated US$45 million will be required to cover 1993 operation costs.[23] Although no reliable data exist for costs of specific inspections, UNSCOM 31, a missile inspection team of 29 inspectors and 6 support personal undertaken in March 21-29, 1992, reportedly cost over $400,000.[24] Thus a *rough* estimate of the cost of the 16 missile inspections conducted to date would be on the order of $7.2 million (16 x $400,000).

Until resolution 778 was passed, funding came from several sources, primarily contributions in cash and in kind from member states.[25] Cash contributions as of June 1992 were the United States ($14 million), Saudi Arabia ($1.73 million), Japan ($1 million), Kuwait ($1 million) and the United Kingdom ($170,000). The UN also contributed eight million dollars from its Working Capital Fund. A wide variety of equipment--vehicles, chemical decontamination gear, communications gear and aircraft--were either donated, or provided on loan to UNSCOM by member states.

Personnel

Salaries of UNSCOM personnel on loan from member states do not pass through the UNSCOM budget, which has significantly reduced the organization's resource burden. Of the 137 full-time positions distributed amongst the headquarters, Bahrain, Baghdad and International Atomic Energy Agency (IAEA) offices in October 1992, 48 positions were financed by the UNSCOM budget (half of which were support staff) with the balance of personnel on loan from governments for assignments ranging from 3-12 months.[26] This pattern of personnel on loan versus UN employees continues today. The UN does pay subsistence allowance and travel expenses for all employees working for the Commission. Currently, the one member of the Assessment Unit ballistic missile team is a permanent UN employee, while the other member is on loan from a member state.

According to IAU analysts, personnel selection, both for the permanent staff and for inspectors, is generally ad-hoc. To identify someone for a particular task, an inspection team member or headquarters employee may suggest someone with whom they have worked. The individual is then contacted informally to determine if they are willing to serve on a team. Only then is a formal letter submitted to a member state by Chairman Ekeus to ask for the individual's services. Naturally this has lead to a preponderance of inspectors or analysts from industrialized or military powers, not only because they have significant technical expertise but also because they are the ones current team members are already familiar with. Primary representation in the IAU and on inspection teams is from the United States, the United Kingdom, France, Australia, Russia, New Zealand, Germany and Russia. In all, 38 different nationalities and 772 individuals have gone on in-country inspections.

Significance

Obtaining dependable sources of income has been a continuing problem for UNSCOM. On several occasions the organization has had to adjust operations to when income would be forthcoming. Any multilateral inspectorate would clearly need an approved, dependable budget in order to successfully plan for its regular operations.

ZBM Implications

Annual budgets for the OSI portion of a UN missile inspectorate would depend partly on inspection modalities (number and type) and the number of countries to be inspected. To obtain a very rough estimate based on the UNSCOM experience, we assume 27 countries requiring regular inspections[27] on the order of 8 inspections per year. To err on the conservative side, we assume another 10 states with a well-developed infrastructure potentially capable of developing a missile system; such states might require one or two inspections per year. Given these assumptions, on-site inspections for a ZBM might total approximately $94 million per year. Of course, these costs would increase as candidate ballistic missile-capable or SLV program states might increase in the future.

In order to ensure continued financing, the General Assembly would have to be involved in approving the missile inspectorate's budget. At a time of financial scarcity, $90-100 million per year--which does not include administrative and other non-OSI costs-- might be difficult for the UN to absorb. Helping to prepare and subsequently lobbying for an annual budget would be a fitting task for the formal inspectorate "commission." Similar to IAEA practices, a fixed annual budget would not preclude the loan of personnel or equipment by member states. Such temporary loans would certainly be required for certain specialized tasks and would help to make up any shortfalls in annual budgets.

As to personnel selection, the ad-hoc nature of the UNSCOM process very likely could not be continued. In its place, one former UNSCOM expert suggested to the author establishment of a kind of "register" of inspectors, made available by member states and subject to the UN Inspectorate's approval, to regularize and make more transparent the personnel selection process.

Intelligence: The Role of the Information Assessment Unit

Given the sheer scope of the *687* mandate, the need for a body within UNSCOM to perform an intelligence function--the collection, analysis and dissemination of information--seems obvious. But due to the initial time-sensitive priority of getting inspectors on the ground, both to establish an on-site precedent and to quickly obtain data on Iraqi weapons programs, UNSCOM paid little attention to the systematic collection and analysis of inspection and other data the Commission was receiving.[28] It was not until the influx of information began to overwhelm the organization, and important documents could not be quickly retrieved and assessed, that the need for an intelligence unit became apparent. It would also appear that the political difficulties of establishing a UN bureaucracy with an intelligence function--with implications of secrecy anathema to the UN structure--were a significant impediment to the establishment of the Assessment Unit itself. The IAU now provides, among other services, the systematic storage, retrieval and assessment of information; assistance in planning and execution of inspections; maintenance of various databases on Iraq's supplier networks; and targeting for surveillance flights. To better assess that role, it is useful to see how the IAU performs the functions of a classic intelligence agency.

The IAU collects data from a wide variety of sources, foremost among them U-2 and helicopter overflights,[29] member state intelligence agencies, member state provision of information based on UNSCOM requests,[30] on-site inspections, Iraqi written declarations, question and answer sessions with Iraqi weapon designers, engineers and officials, and from the public media. Perhaps most interesting is the collection and use of data provided by member state intelligence agencies. Particularly during the first 18 months of inspections, such intelligence was plentiful and of a high quality as a result of the massive data gathering effort for the Gulf War. Member state intelligence included, in many cases complete evidence, assessment, conclusion and ground coordinate packages. The United States has been the primary provider of intelligence to UNSCOM, but important contributions have also come from the United Kingdom, Russia, France, Germany and Ukraine.[31] However, the intelligence well from member states is now "getting drier and drier" according to one UNSCOM official, and the IAU has begun to have strong fears of "harassment intelligence" being provided. Thus, the IAU is now relying more heavily on leads (inadvertently) provided by the Iraqis themselves and by systematically developing the data reported in the public media, particularly in terms of Iraq's supplier network.

Initially, UNSCOM relied heavily on member state (particularly US) analysis of data collected during inspections. But with the establishment of the IAU, UNSCOM acquired a capability to do in-house analysis of such data. Due to the emphasis on actionable intelligence and to the paucity of available analysts, no formal, conclusive assessments of the Iraqi missile program have been written. With only two analysts working on missile issues, the IAU has simply not had the time nor the resources to do more complete assessments, and inspection support--such as establishing priorities for site inspections--has tended to be the priority. Aside from inspection support, analyses of supplier networks, the inter-connection between nuclear and missile programs and between various military-industrial organizations, and of the completeness of Iraqi declarations, are also important aspects of the IAU's work.

The IAU does not write for a wide audience; the Operations Division and the Chairman are the primary recipients of the unit's intelligence analysis. Official written records, such as the semi-annual reports on UNSCOM operations that Chairman Ekeus is required to provide to the Security Council (and thus to the UN in general) are drafted by the Political Advisor. However, it should be noted that an executive summary of each inspection is submitted to the Executive Office of the Secretary General, along with a transmittal memo noting that "You may wish to make the attached executive summary available to members of the Security Council as well as to other interested Member States."[32] The summaries are not very detailed nor are they publicly available, but states interested in the reports do have access to them. It appears that not all member states have access to the complete, detailed inspection reports. As to the raw intelligence data, an informal quid pro quo relationship was worked out in at least one case between UNSCOM and two member states providing significant logistical support for the inspections: in exchange for that support, the member states expected to get first-hand "ground-truth' information from the inspectors just as they came out of Iraq.[33]

Significance

As can be seen from the discussion above, the IAU has responsibilities analogous to a typical national intelligence agency. Indeed one UNSCOM official noted that the IAU is "the first intelligence unit ever within the UN."

Similar to the establishment of UNSCOM itself, the greatest significance of the IAU may be simply that it exists and has proven that some form of an intelligence unit can function within an international organization.

However, in many ways the IAU does not act as a classic intelligence agency and given the cultural and political milieu within which the organizations operates, this outcome is only to be expected. There is no formal compartmentalization of data within the IAU or UNSCOM although say, the biological weapons analysts, generally do not ask to see all the data in the hands of the ballistic missile analysts. Collector and analyst are often one and the same person, which *may* lead to pathologies in the information process; the analyst may add additional weight to a piece of evidence provided by an Iraqi official with whom he is familiar or from a particular site with which he has dealt with on a long-term basis. Moreover, the security of information and facilities in no way approach what one might inspect of an "intelligence agency". Nonetheless, simply having punch locks on doors has tended to separate UNSCOM from the rest of the UN bureaucracy.

Several intelligence problems are unique to UNSCOM, including analyst nationalities; sharing of member state data within an international organization; and the subsequent use of that data to make verification and compliance decisions. Analysts have had to realize that they are indeed working for the UN and not their own country. This boundary is difficult to define and maintain as most of the IAU's analysts have close national or military intelligence connections. Credibility of analysis is here at stake, for if member states perceive that analysts are working for their own country rather than the UN, then the data they produce will not be taken into account.

The IAU has "solved" the problem of sharing intelligence data within the organization by making it clear to member states that the Chief of the IAU will share that data with the Chairman and other UNSCOM officials. As one official told the author, an international intelligence G-2 can never say to his boss, "I can't tell you where I got this information." According to the official, such compartmentalization simply would not work within the UN milieu. Perhaps due to the tact with which UNSCOM has handled intelligence, non-compartmentalization of data has not lead to any significant problems with the flow of information into UNSCOM.

ZBM Implications

Despite several shortcomings, the Information Assessment Unit would generally be a good model for a ZBM. Several obvious changes would need to be made. First, due to credibility problems, it is obvious that any *quid pro quo* intelligence sharing relationship could not be allowed to exist. A ZBM treaty could be negotiated only among partners who must feel they are not being discriminated against by virtue of their participation. Second, the security of facilities and intelligence holdings must be enhanced to provide member states assurances that their information will not be compromised.

There is a further problem: in order to make compliance decisions on a ZBM treaty, the Security Council would obviously need to have access to the

inspectorate's evidence and assessments. But simply allowing for a wider audience to see assessments will introduce a perplexing dichotomy. To the extent that a ZBM equivalent of the Assessment Unit relies on member state intelligence: assessments and data must be both credible, that is, it must have a certain specificity; and it must be sharable, that is, Security Council and perhaps other officials must be allowed to see it. Given the national intelligence agencies concern to protect and keep secret their sources and methods, one may expect that sharing and specificity of intelligence data are likely to be inversely related.[34]

One way to solve many of these problems would be to regularize the collection of missile-related intelligence from member states.[35] The UN charter provides an institutional mechanism for "harnessing national intelligence efforts to international objectives", namely, the Military Staff Committee. The Military Staff Committee has the power to create subsidiary bodies. In this particular case, a Joint Intelligence Staff could be created to serve as a focal point for screening national intelligence provided to the international body as well as serve as a useful means to protect national sources and methods.

Inspections, Interim Monitoring and Ongoing Monitoring and Verification

Implementation of *687* provisions with respect to Iraq's ballistic missile capabilities (and to its WMD capabilities in general) have proceeded along lines analogous to those implemented under the US-Soviet INF Treaty. Indeed, several of UNSCOM's missile analysts and inspectors were previously INF on-site inspectors. Briefly, implementation entails an inter-related, three stage process of: first, an inspection and survey phase, based initially on Iraqi declarations and subsequently from data gathered by UNSCOM itself, which was intended to establish an informed assessment of Iraqi capabilities and facilities in the missile field; second, the disposal or rendering harmless of all *687*-prohibited systems, facilities and equipment; and third, the long-term monitoring phase to ensure Iraqi compliance with *687* obligations not to reacquire prohibited missile capabilities. In practice, there has been little distinction between the first two phases, as much destruction of missile equipment was undertaken with the first inspections.[36]

The OMV Plan[37]

UNSCOM's OMV plan, approved by the Council in its *Resolution 715* of October 11, 1991, consists of several key elements:

- *Inspection*: Under the plan, UNSCOM maintains the right to inspect any "site, facility, activity, material or other item in Iraq" without hinderance and, if it so chooses, without notice. In other words, near complete intrusiveness.

- *Iraqi obligations*: Iraq is required to provide regular reports of information on "activities, sites, facilities, material and other items, both military and civilian, that might be used" for 687 prohibited activities and to provide information or clarifications as requested by the Special Commission. Iraq must also allow unconditional fixed and rotary-wing overflights throughout the country.

- *National implementation measures*: Iraq must adopt administrative and legal measures which will prohibit all persons acting under Iraqi jurisdiction from "undertaking anywhere any activity" prohibited by 687.

- *Import control regime*: The plan calls for an import control mechanism that "provides for timely information about any sale or supply to Iraq for items that could be used not only for permitted purposes but also for the purposes prohibited under *Resolution 687*." The plan contains detailed technical and equipment annexes which fall in this category. For missiles, the list is very similar to the MTCR annexes. It is important to note that the provision is to remain in effect even after the embargo against dual-use equipment sales is lifted. UNSCOM intends for the Council's Sanction Committee to evolve into an arms embargo and licensing committee for dual-use items, while the Special Commission and the IAEA would be responsible for monitoring an item's use (including spot checks and challenge inspections) once it has arrived inside Iraq.[38]

- *Non-compliance*: In the event that UNSCOM discovers any prohibited activities, facilities or equipment, the Commission "shall have the right to take it into custody and shall provide for its disposal as appropriate," with Iraq retaining no ownership rights to items to be destroyed. Of course, the Commission also retains the right to raise compliance issues with the Council.

With respect to missiles, all items laid out in the technical and equipment annexes are subject to the verification process. Iraq is required to present to the Commission, *inter alia*, a list of all surface-to-surface missiles for use or capable of being modified for use with a range greater than 50 km, and to specify the name, type, type of propulsion, guidance system and airframe, sites or facilities where located, launcher types etc. Iraq must also provide detailed information

on any dual-use equipment imports, including supplier and quantity, point and time of entry, end-use facility and name of importing organization.

The strategy behind the OMV plan generally mirrors that of any viable and effective verification regime: to assure high confidence that militarily significant cheating will be detected in a timely manner. But the UNSCOM plan, according to officials and analysts, adds a subtle political objective as well: that of changing Iraqi intentions with respect to the WMD and delivery system programs. UNSCOM proposes to establish a monitoring regime that is so powerful that it convinces the regime that calculations of Iraqi power based on the maintenance of WMD programs and capabilities cannot succeed since any reconstitution of those capabilities will be detected by UNSCOM.

At the time of this writing, detailed and routinized OMV procedures have not been worked out within UNSCOM. General outlines of the plan do exist and, according to one official, UNSCOM could begin to implement it at short notice. Again the question of resources and priorities have intruded on the Commission's and in particular, the IAU's ability to present a detailed plan. To date, a lot of work on routinizing the plan has been done by individual analysts, but not in a comprehensive fashion.

Organizationally, UNSCOM is expected to change little with the onset of the OMV phase, except for the addition of one or two IAU analysts who will work on specifically on the import control regime.

The onset of the OMV regime awaits Iraq's written acceptance of the provisions of Resolution 707 and Resolution 715. The former requires full, final and complete disclosure of all activities related to WMD programs while the latter requires Iraq to accept all conditions and obligations of the OMV plan. While Baghdad argues that acceptance of *715* obligations would infringe on its national security and national sovereignty and has repeatedly claimed that it has fulfilled all its *687* obligations, UNSCOM and the Security Council maintain that Iraq has not fulfilled the conditions. According to Rolf Ekeus, "Should the Commission now seek to initiate the ongoing monitoring and verification phase of its mandate under these circumstances, it will be sending a message that, in fact if not in law, it is prepared to operate ...under Iraq's, not the Council's conditions. Past experience has demonstrated the very serious inadequacies of such an approach."[39] The two sides have been at loggerheads for well over a year and a half on this issue.

Interim Monitoring

In order to ensure that there are no Iraqi attempts to re-establish its missile programs prior to onset of the formal OMV regime, UNSCOM has begun what it terms Interim Monitoring and Verification. Beginning on January 25, 1993 and continuing for 60 days thereafter, UNSCOM has maintained at least one of three inspectors in an around the clock vigil at Iraq's primary missile R&D center located at Ibn al-Haytham. The inspectors have had what can be described only as total access to the facility including "on demand" access to any office, desk or computer file as well as the right to conduct lengthy interviews with al-Haytham's engineers and officials. Missile experts now believe they have a full knowledge base of the site, including workers schedules, the current status of prototype designs and the center's relationship to the rest of the missile and military industrial bureaucracy. Since the onset of the IMV, its scope has been expanded to include five primary and six secondary sites (al-Haytham is now visited approximately once a week), with the focus on closely monitoring guidance and control R&D and solid rocket motor production.

Significance

It is difficult at this point to assess the significance of the OMV plan since it awaits full implementation. It is useful however, to note some of the plan's more interesting characteristics. Perhaps the most striking feature is the level of intrusiveness it calls for. Indeed, the entire process of disarming Iraq has been characterized by this high degree of intrusiveness. With the OMV, UNSCOM, backed by explicit Security Council authorization, maintains the right of unrestricted access to any site, the right to request and retain data, to conduct interviews with relevant Iraqi officials and weapon specialists, to install monitoring equipment, to collect samples, and to conduct aerial surveillance of suspected sites. Of course, there have and will continue to be Iraqi efforts to obstruct or restrict these rights. But by and large the inspection teams have had unprecedented access to military and civilian production facilities in what is, after all, still a sovereign state. International political will has afforded this degree of intrusiveness, but it is far from clear--once the Iraqi question has been eclipsed by more pressing international problems--that it can be successfully maintained in the years to come.

Another important feature is the synergy of the various verification instruments and obligations. The Council and UNSCOM have put forth a plan which approaches the monitoring problem from both sides of the issue. For example, the import control regime obviously serves to address Iraqi attempts to acquire foreign technology for its prohibited programs, while Iraq's administrative obligations (prohibitions against Iraqi personnel undertaking 687 banned activities) explicitly lays the onus of responsibility on the Iraqi government for the activities of the vast reserve of military-scientific expertise

that remains in the country. At the very least, the synergy serves as a strong disincentive for Baghdad to covertly repudiate its 687 and 715 obligations.

ZBM Implications

Given that *Resolution 715* was adopted under Chapter VII of the UN charter, all member states are required to assist in the implementation of the OMV plan. Thus they are required to implement obligations not to export prohibited items to Iraq. But a ZBM regime would not be so stringent, as only states party to the treaty would be legally bound by the terms of the agreement. Also, it would also be highly unlikely that UNSCOM's right to take immediate possession or to destroy banned equipment would appear in a negotiated agreement. In the ZBM case, the ultimate disposition of questionable technologies obviously would be subject to an established dispute resolution mechanism. And finally, the inspection modalities would need to be agreed to by all parties prior to assumption of treaty obligations. Making the modalities more transparent might prove to be the job of an equivalent body of Commissioners.

The OMV plan does offer some useful lessons for a ZBM. In particular, the import control regime could serve as a model for a Space Technology Registry, wherein states party to the treaty would undertake to advise the monitoring and verification agency of any sales of treaty relevant technologies. Towards this end, the assessment unit (much like UNSCOM) would require analysts and experts dealing with covert procurement issues as a confidence-building hedge against any illegal activities. Meanwhile, the spirit of Iraq's explicit obligations under the OMV would remain implicit in the signing of the treaty as states would be legally bound to prohibit nationals from engaging in banned activities.

Conclusions and Study Recommendations

Several general conclusions result from the above study. First, UNSCOM is a unique organization borne of a unique set of circumstances and it is therefore unlikely to be repeated. Iraqi disarmament is the result of a ceasefire accord. The terms of any negotiated nonproliferation agreement will likely be vastly different as a result of this simple fact. As such, any attempt to extend the responsibilities of the Special Commission on Iraq, which is essentially a single issue organization, would likely result in failure. Of course, this difference does not preclude learning lessons from the UNSCOM experience nor the creation of a separate UNSCOM-like body within the United Nations. Most important in this regard, would be to maintain UNSCOM's small and efficient organizational and decision-making structure.

Second, any establishment of a UN missile monitoring and verification organization should be as a subsidiary body of the Security Council. Time and again, the unwavering support of the Security Council has proven invaluable in UNSCOM's successful efforts to carry out its responsibilities.

Third, it has now been established that a monitoring and verification agency can be viable within the UN context. When given a clear mandate and when provided with the necessary resources and freedom from UN's legendary bureaucratic processes, a diverse body of international inspectors and analysts can accomplish difficult nonproliferation and disarmament tasks. Moreover, UNSCOM has established that a UN intelligence or information agency (the Information Assessment Unit) can receive, analyze and act upon information received from member states and other sources in a credible manner.

In this study I have only scratched the surface of UNSCOM's significance and implications for future missile nonproliferation regimes. More work needs to be done on the various operational issues: how to distinguish "peaceful" space technologies from "military" missile technologies or how to establish international inspection modalities that provide a high degree of verification confidence among interested parties. From the political perspective, research must continue on the changing role of the United Nations and the degree to which the organization can absorb new and difficult tasks of monitoring and verifying nonproliferation agreements.

Notes

1. This study was written prior to the author's appointment as an UNSCOM ballistic missile inspector. The author would like to thank Catherine P. Dalipe for her assistance in preparation of this study.

2. Two useful studies in this regard are Frank R. Cleminson, "United Nations Special Commission on Iraq Pursuant to SCR 687 (1991): Verification of Future Compliance," in Serge Ser, ed., *Verification of Disarmament or Limitations of Armaments: Instruments, Negotiations, Proposals*, United Nations Institute for Disarmament Research (New York, 1992), pp. 253-261; and Johan Molander, "The United Nations and the Elimination of Iraq's Weapons of Mass Destruction: The Implementation of a Cease-fire Condition," in Fred Tanner, ed., *From Versailles to Baghdad: Post-War Armament Control of Defeated States* (United Nations Institute for Disarmament Research, 1992), pp. 137-157.

3. For the purposes of this study, monitoring is defined as the "process of watching, observing or checking objects, activities or events, for a specific purpose," and verification as a "process which establishes whether the States parties are complying with their obligations under an agreement." UN Department for Disarmament Affairs, *Study on the Role of the United Nations in the Field of Verification*, Report of the Secretary General (United Nations: New York, 1991), p. 4.

4. For a more thorough examination of these issues, see chapter 11.

5. Of course, the proposal is much more detailed and nuanced than what I have allowed for above. For a complete description of the proposal, see Lora Lumpe, "Zero Ballistic Missiles and the Third World," Center for International Security at Maryland, Project on Rethinking Arms Control, Paper No. 3, March 1993; and chapter 9.

6. This assumption is emphasized as a"point of departure" in *Study on the Role of the United Nations in the Field of Verification*.

7. For a more detailed discussion of the technical and other difficulties with a SLV safeguard regime, similar in many respects to the ZBM proposal, see Brian Chow, *Emerging National Space Launch Programs: Economics and Safeguards*, R-4179-USDP, RAND, 1993, pp. 55-65. See also Jurgen Altmann, "Ballistic Missile Limitations and Their Verification," in Gotz Neuneck and Otfried Ischebeck, eds., *Missile Proliferation, Missile Defense and Arms Control*, Institute for Peace Research and Security Policy, University of Hamburg, 1993, pp. 91-99.

8. George Bunn, Draft Memo to the International Organizations and Nonproliferation Project, Monterey Institute of International Studies, "What does

the UN Security Council Order Directing Iraq to Disarm *(Resolution 687)* Stand for as a Precedent?," May 15, 1993, p.6, fn. 8. The following paragraphs draw heavily on Ambassador Bunn's analysis.

9. Rolf Ekeus, "The Iraqi Experience and the Future of Nuclear Nonproliferation," *The Washington Quarterly,* Autumn 1992, p. 68.

10. Interestingly, while 687 specifically uses the term "ballistic missiles" the OMV plan notes that "The prohibition applies to any ballistic missiles *or missile delivery systems."* See *Plan for Future Ongoing Monitoring and Verification of Iraq's Compliance with Relevant Parts of Section C of Security Council Resolution 687 (1991),* S/22871/Rev. 1, October 2, 1991, para. 41 (emphasis added).

11. See UN Security Council Provisional Verbatim for April 3, 1991, S/PV, 2981. References were made, however, to prior council resolutions with respect to Iraq.

12. Boutros Boutros-Ghali, *New Dimensions of Arms Regulation and Disarmament in the Post-Cold War Era* (UN, Department of Political Affairs, January 1993), p. 8.

13. The UN's Disarmament Yearbook stated that the circumstance's surrounding 687 "are, in many respects unique and therefore the experiences do not necessarily serve as guidelines for the future. On the other hand, there can be little doubt that important precedents have been set that vividly demonstrate the extent of multilateral actions that can be taken by the international community in concert in the face of threats to international peace and security." United Nations, *The United Nations Disarmament Yearbook, 1991* (UN Office of Disarmament Affairs, New York, 1992), v. 16, pp. 53-54.

14. United Nations, *New Realities: Disarmament, Peace-building and Global Security,* Excerpts from the panel discussions organized by the NGO Committee on Disarmament, April 20-23, 1993 (New York: United Nations, 1993), p. 261.

15. Article 39, Chapter 7 (enforcement provisions) of the UN charter states that "The Security Council shall determine the existence of any threat to the peace, breach of the peace, or act of aggression and shall make recommendations...to maintain or restore international peace and stability."

16. See S/PV.3046, January 31, 1992.

17. In his report on the Special Commission's formation, the Secretary General stated that, "I wish to emphasize the need for an efficient and effective executive body." The Secretary General clearly seemed to see that the Commission would not be able to take a lot of time and do things the "UN way."

See *Report of the Secretary General: Implementation of paragraph 9 (b)(i) of Security Council Resolution 687* (1991), S/22508, April 18, 1991.

18. Initial work on the Assessment Unit's formation began in October 1991.

19. Pierce S. Corden, UNSCOM Deputy Executive Chairman, communication with the International Organizations and Nonproliferation Project, Monterey Institute of International Studies, March 11, 1993.

20. This figure does not include the hundreds of inspectors who are usually with UNSCOM only during inspection and pre- and post- inspection briefings.

21. It should be noted however, that the CBW Working Group, along with the Destruction Advisory Panel, did considerable planning for the destruction of Iraqi chemical munitions.

22. See Report by the Executive Chairman of the Special Commission, S/23165, October 25, 1991, para. 25-31.

23. Corden, communication with MIIS; and UNSCOM documents provided to the author. The 1993 costs include estimated IAEA expenditures for the removal of irradiated fuel from Iraq and for its permanent disposal.

24. GAUCHER and National Security Research, Inc., *Iraq Inspections: Lessons Learned*, Final Draft Technical Report for the Defense Nuclear Agency, DNA001-91-C-0030, September 11, 1992, p. 97.

25. See S/24108, Appendix VI.

26. UNSCOM documents provided to author.

27. This number includes the five recognized nuclear powers plus 22 emerging missile capable states. See "Missile Capabilities of Selected Countries," *Missile Monitor* (No. 3, Spring 1993) International Missile Proliferation Project, Monterey Institute of International Studies, pp. 1-4.

28. For example, until the IAU's formation, there was no central repository for all the inspection reports.

29. The US government provides the U-2 to UNSCOM. The pilot flies with UN identification papers and the IAU determines the specific sites to be photographed.

30. For example, at the early stages of operations, the Commission invited member states to provide information on relevant equipment and technologies exported to Iraq prior to the war. However, a relatively small number of countries responded.

31. Germany has provided a vast amount of missile supplier network data, while Ukraine hosted one ballistic missile analyst to study Soviet-style missile production techniques in order to better understand the Iraqi manufacturing

process. Interestingly, early in 1992, Iranian military officers were asked to supply data on Iraqi missile activities during the Iran-Iraq war. Author discussions with UNSCOM, UN and Ukrainian officials.

32. Documents provided to the author.

33. This interesting piece of information was provided by one former inspector, although other UNSCOM officials and analysts would not comment on the matter. It is unclear if this informal quid pro quo relationship continues to exist today.

34. This problem may not be as big as it first appears. As a state party to the treaty deems a particular compliance issue important, it will provide a quality of data or intelligence consequent with that perceived importance. In other words, the risks associated with potential exposure of sources and methods should be weighed rationally against the significance of releasing that information. For example, the US Central Intelligence Agency (CIA) provided important information to the IAEA on the North Korean nuclear facilities at Yongbyon. Although downgraded, the information was specific enough to cause the IAEA to act upon it.

35. The following is based on David Kay, "Nuclear Proliferation in the 1990s: Challenges and Opportunities," Speech for delivery at the Woodrow Wilson Center, December 1-2, 1992.

36. See United Nations Department of Public Information, "United Nations Special Commission," (Advance Copy), DPI/1239/Rev. 2, April 1993; and Plans for the Implementation of Relevant Parts of Section C of Security Council Resolution 687 (1991), S/22614, May 17, 1991, arts. 5-26.

37. The plan is detailed in S/22871/Rev. 1, October 2, 1991.

38. Discussions with UNSCOM officials and documents provided to the author.

39. *Special Report by the Executive Chairman of the Special Commission Established by the Secretary-General Pursuant to Paragraph 9(b)(i) of Security Council Resolution 687 (1991)*, S/23606, February 18, 1992, para. 16. For a more complete discussion of this dispute see Tim Trevan, "UNSCOM Faces Entirely New Verification Challenges in Iraq," *Arms Control Today*, April 1993, pp. 11-15.

11

Verification of
Ballistic Missile Activities:
Problems and Possible Solutions

Peter D. Zimmerman[1]

A nation need not take home the gold medal in the military-technological Olympics in order to acquire a stockpile of weapons of mass destruction and guided missiles for their delivery. It can strive merely for the bronze medal, using low-tech methods, in order to obtain an arsenal which can deter or compel its neighbors with the threat of nuclear destruction, delivered by guided missile. It is preferable for an aspiring missile power to use "bronze medal technologies"--those which have been tried and proven, which are mature and described in detail in text books and taught in engineering colleges throughout the world, which are components of commercial products--than to try to emulate the most advanced products of the developed nations, and fail in the attempt.[2] A bronze medal earns a proliferant state a place on the winners' podium.[3] This was certainly the case in Iraq and, to an astounding extent, in South Africa where a technically advanced nation used minimal technology to produce easily built and reliable nuclear weapons.

Beginning with a national decision to develop nuclear weapons in 1974, the Republic of South Africa fuelled its first air-deliverable weapon in early 1978. The South African design was of a gun-type weapon, undoubtedly overdesigned, since it was only 1.8 meters long and 65cm in diameter, weighing "about a ton." The device was built *without* a neutron initiator, because its designers were little interested in a predictable yield. South Africa operated its uranium enrichment facility with 0.5 percent tailings during the early years of its weapons project, and loaded its first weapon with uranium with an enrichment in the vicinity of 80 percent ^{235}U rather than the 90^+ percent which they used later on. The South African program, which cost only 700-800 million Rand (less than $300 million), never employed more than about 400 people at a time, and only 1,000 worked on the program over its entire 16 year history.[4]/

The inability to deny developing nations the ability to construct bronze medal weapons indigenously naturally leads to consideration of proliferation management regimes, non-proliferation regimes, and, in today's fashionable term, "counter-proliferation regimes." For such regimes to be effective, they must offer advantages to potential proliferant states as well as to states wishing to constrain the armaments of others. In short, the appropriate methods to control proliferation involve carrots, not only sticks; require mutually satisfactory agreements between potential supplier states and potential proliferants; and imply treaties, not "control regimes." Treaties, to be satisfactory, must be verified, and means for such verification with respect to "bronze medal" missile technologies will be discussed in this paper.

Technical Background to Missile Proliferation

We are normally encouraged to view long range guided missiles and nuclear weapons as developments requiring the best of first world technology to design and produce. In fact, neither missiles nor the warheads to fit atop them require 1990s technology, nor even that of the 1970s. Both types of weapons were fully mature before 1960, and both entered military service in 1944 (the V-2) or 1945 (first generation fission weapons). Because guided missiles and nuclear weapons which can pose both regional and trans-regional security problems can be built using essentially obsolete technology, it may be difficult to monitor the development of indigenous capacities in developing countries simply by observing the import practices of the target nations. A brief review of the ease with which missiles can be built is in order.

The threshold range for missiles covered by the Missile Technology Control Regime (MTCR) is 300km; the threshold payload for missiles controlled by the regime was 500kg, but has since been reduced to "zero" in recognition of the fact that biological weapon payloads can be very small. The German A4 (V-2 in service) rocket exceeded the old MTCR payload limit by almost a factor of two and matched the range limit. With a 750kg warhead, the A4 would have had a range significantly in excess of the 300km. Its first successful flight was on October 2, *1942*. More than 2,000 A4 missiles were used in combat, and close to 80 percent were successful.

To understand the problems a proliferator might encounter in designing and building a third world missile, the model to study is the A4, not the modern American Pershing II.

Cruise missiles are older technology. The Fieseler Fi-103, a pulse-jet powered cruise missile saw service as the V1 "buzz bomb" on the German side of World War II from mid-1944 onwards carrying a payload of 1,000kg of high explosive about 200km (some Fi-103s were more efficient or carried greater fuel loads and are reported to have had ranges of 300km). If integrated circuit chips were substituted for the bulky electronics required by 1940s technology the fuel load of the Fi-103 could have been increased significantly, lengthening its range. Neither the A4 nor the Fi-103 was particularly accurate[5], but even compared to modern missiles they were tolerably reliable.[6] Neither missile, of course, used any semiconductor electronics, relying instead on the only thing which was available: tubes and relays. The simple substitution of guidance systems using solid state electronics would permit a significant increase in range and payload, (tube electronics are heavy) and accuracy, because of the vastly increased computer power available. The reliability of the control systems, and hence of the missiles themselves, would also improve.

Engineering drawings for both the A4 and the Fi-103 are readily available. Both the Smithsonian Institution in Washington, DC and the *Deutsches Museum* in Munich appear to have complete sets as well as operating manuals, service and workshop manuals, and the other documentation needed to reconstruct substantially similar airframes and engines. Of course, no one would use as crude and inefficient a propulsion system as the pulse jet on the Fi-103 today; far better turbojet engines are available off the shelf in any aircraft maintenance station, and are built in countries with as low a state of industrial development as Turkey.

Although supercomputers are useful to the most technologically advanced nations for developing new missiles and new nuclear weapons, such computational power dwarfs the ranks of hand-operated adding machines used by rows of young women to perform the numerical analysis needed to develop the first nuclear weapons. Indeed, desk-top computers vastly exceed the computing power of the most advanced mainframes in use at the time of the design of the Atlas and Titan intercontinental ballistic missiles and their warheads.

Should a potential proliferator seek missiles with longer ranges and greater payloads, the Smithsonian Archives in Suitland, MD, have the (almost) complete blueprints for the Jupiter intermediate range missile of the 1950s (2,400km range, payload about 1 tonne). The plans are contained on more than 500 reels of microfilm, but the museum will sell any interested scholar duplicate copies for about $11.50 per reel. Because the technology of the Jupiter is considered to be so obsolete, the blueprints are not subject to export controls.

To be sure, the available Jupiter technology is *not* complete; details of the guidance electronics are unavailable. But a modern general-purpose laptop computer, using perhaps a 486-SX processor chip, is far more powerful than the specialized computer of the Jupiter; the laptop is also quite rugged. The nuclear warhead is also not described, but a significant quantity of information about the heatshield for the reentry vehicle (RV) will be found in the microfilm. *All of the technology needed to design and build ballistic and cruise missiles with ranges on the order of 300km is publicly available for the asking. Much of the technology to build 2,400km range ballistic missiles is also in the public domain.* Nobody would copy an A4 or a Fi-103 from its blueprints, nor would they be likely to attempt to reproduce the considerably more complicated Jupiter from the microfilm. The importance of these documents lies in their ability to show a competent designer who is having some difficulty with a particular point how the pioneer researchers solved the problem. It is far easier to be the second (or the tenth) to attack a problem, when it is known that a solution does exist, than it is to be the first to attempt the work.

"Technology" for a missile or nuclear weapons project does not merely exist on paper as plans and equations. It also requires hardware, and the MTCR attempts to tend to the hardware side of the problem. As Iraq has shown us, however, much of the critical hardware for constructing rockets and their warheads has other purposes as well. Computer numerically controlled machine tools are the mainstay of modern heavy industry; one way or another, they can be bought. Vacuum pumps and seals are largely commodities as well. In the nuclear field maraging steels and carbon fiber rotors can be had somewhere. When there is as much money involved as, for example, Iraq had to spend, willing buyers and willing sellers of illicit products will find one another, even if they must do so through shady middlemen. Despite the marginal legality of the market, Iraq paid only a few cents a pound "risk premium" for its maraging steel.[7]

Indeed, most of the relevant technology for production of V-2- or Scud-like missiles already exists in developing countries, although not all of the technology is likely to be found in any one nation. However, the commercial network to connect third world suppliers and consumers of the equivalent of late 1950s American or European technology is well developed, so lateral transfers should be a tractable problem.

The MTCR and "Supply Side" Controls

The MTCR assumes that if first world supplies are choked off, then proliferators cannot readily construct adequate guided missiles. In fact, the A4 is a perfectly adequate terror weapon, and with a nuclear warhead of 1,000kg mass (about what Iraq would have finally built, according to knowledgeable sources), a fine strategic weapon.

Fundamentals of "Supply Side" Controls

The fundamental assumptions of the MTCR are:

- There is a core of controlled and controllable technology.
- Without access to the core of controlled technologies, nations of the developing world cannot develop missiles indigenously.
- The cost of circumventing the control regime is enough to deter virtually all problem countries from even making the effort.

Failure of Supply-Side Controls at the "Bronze Medal Level"

Iraq has proved that all three of these assumptions are false. Virtually all of the technology needed to construct liquid propellant and solid propellant missiles with ranges of less than 500km can be obtained on the open market or developed indigenously by any nation with an airframe industry, access to common industrial chemicals (such as nitric acid and hydrazine) and the equipment to handle them in safety, and the ability to repair jet and turboprop engines. Most countries which are likely proliferants have at least these capabilities. It is, therefore, difficult to prevent the proliferation of missiles with ranges under 500km by supply-side controls. The controlled materials are, in fact, ubiquitous, as is the knowledge of how to assemble the raw materials into a workable missile. It is not unreasonable to contemplate a certain relaxation of export controls as they pertain to missiles in this range class while simultaneously pursuing a "demand side" policy in non-proliferation. Such a policy would include security guarantees against missile attack and also enthusiastic assistance, where requested, in providing the targeted nations with access to space for peaceful uses.

However, as the range of a ballistic missile increases, the difficulties of construction also increase, literally, exponentially. The designer is forced to squeeze every gram of superfluous mass out of the airframe and the propulsion and guidance units. The Atlas missile, the first American ICBM, still used as a space booster, cannot even stand up unless its propellant tanks are pressurized because its skin is so thin. Indeed, the skin of an Atlas can be pierced by a sharp pencil. The warhead of an intermediate or intercontinental missile must be protected against the stresses and temperatures of reentry, and despite the availability of general information on heatshields of various types, constructing a heatshield with adequate precision is a formidable task. Finally, a guidance system which would permit a 5,000-10,000km range missile to reach a target even as large as the heart of a city remains difficult to construct.

Guidance -- A Limiting Technology

At ICBM ranges the limits on guidance accuracy are severe, particularly if the weapon is designed to strike hardened targets. Even with warhead yields of 100-500 kilotons, CEPs (Circular Error Probable) on the order of 100 meters are required to destroy missile silos with normal hardnesses. Although the SALT I Interim Agreement defined an intercontinental missile as having a range of 5,500km, many have ranges of 10,000km.[8] The maximum useful range for any ballistic missile is 20,000km, half the circumference of the planet.

A CEP of 100 meters at 10,000km requires a precision of one part in 10^5, a truly stressing specification, and one met by the missile designers of at most two nations. A 100 meter CEP at a range of 300km requires a guidance accuracy of only three parts in 10^4, which is significantly easier to achieve.

In fact, however, a 300 meter CEP is more accurate than would be necessary for a plausible bronze medal missile carrying a nuclear weapon with a yield of 20 kilotons, a readily achievable value for a plutonium implosion system. The 5 pounds per square inch (5 psi)[9] radius for a 20 kt surface burst is roughly 1.2km. A CEP of half that value will virtually ensure the destruction of any target specifically selected.

Cruise Missiles

Cruise missiles are inherently easier to build than ballistic missiles, since they consist of a simple airframe and an expendable propulsion unit. Although the very cheap jet engines which made the American ALCM and SLCM so attractive may not be readily available to a proliferator, small jet engines, appropriate for "Executive" jet aircraft can be readily acquired. The modest additional contribution to the total budget for a force of a few dozen cruise missiles made by choosing commercial engines intended for extended use rather than single flights is probably not enough to deter a determined proliferator.

Airframes for cruise missiles need not be developed; missiles can be built simply by converting existing civilian aircraft or pilotless military drones which probably have ranges and payloads well beyond the MTCR limits.

Guidance for a cruise missile is easier to achieve today than ever before. A simple autopilot can attend to the task of keeping the wings level, the throttle correctly set, and the aircraft in level flight at the selected altitude. Navigation of gold medal cruise missiles heretofore has been achieved by the use of complex terrain-matching radar which can fix the position of the missile by comparing digital maps of portions of the flight path with the ground beneath. For all but the most advanced nations, such guidance systems have been overtaken by events, by technologies developed by the United States and the former Soviet

Union. The Global Positioning System (GPS) and *Glonass* satellites of the two nations provide position measurements in latitude, longitude and altitude, updated roughly every two seconds, with an accuracy of no worse than 100 meters. The output of a commercially available Japanese-built GPS receiver (costing less than $500 and readily available around the world) can be fed to the inputs of the autopilot and used to perform precise navigation. Slightly more expensive GPS receivers use signals from up to six satellites simultaneously and automatically calculate correction vectors so that the autopilot need be accurate to no better than a degree over a time as short as a few seconds. To be sure, the precise location of the target must be known so that the GPS system can bring the missile to its goal, but this can be established before conflict breaks out simply by carrying a GPS receiver to the site, or (if the site is in a denied area) to three or four sites bearing a measurable relationship to the target. Alternatively, Landsat and SPOT images, geographically corrected to a few meters absolute accuracy in latitude and longitude, can be purchased for less than $5,000 and are detailed enough to provide targeting information for all but hardened targets which must be attacked with precision munitions. *GPS navigation systems, which are freely available, make cruise missile accuracy independent of the distance flown, both in azimuth and range. One can conclude that cruise missiles can now be built as low risk nuclear delivery vehicles.*

Controlling the Supply Side

The MTCR attempts to control the flow of technology. It rests on strict enforcement by the governments of the supplier nations, controls and limits which must be (but are not) interpreted similarly by all supplier countries and by all of the relevant agencies within the government of each supplier state. Further, supply-side controls must be agreed to by all nations capable of serving as suppliers. Clearly, not all suppliers have agreed, as recent American decisions concerning China's deliveries of missiles to Pakistan demonstrate.

As we have seen in Iraq, the profits available from trade in controlled hardware and software are so great that there is a strong temptation for all but the most honorable of corporations to make the sales and launder the purchases and proceeds. Records, which I have personally inspected, of transactions involving Iraqi front companies such as Matrix Churchill (which appear to be unrelated to the Iraqi government, at least at casual inspection, but which are, in fact, intimately connected to important ministries) and legitimate firms such as Kennametal and Leybold show that many once-respectable corporations have succumbed to the lure of easy money from selling products at prices geared to the special market and its risks.

This situation is reminiscent of the narcotics trade. Supply-side controls have not stemmed the flow of narcotics into any country with which the author is familiar, while draconian demand-side controls employed in (among other nations) Malaysia, Singapore, and Saudi Arabia have significantly reduced the drug problems in those lands. When large amounts of money are at stake, and when the penalties for getting caught can be treated as a discountable cost of doing business, supply side restrictions are likely to be ineffective.

Technology Controls: Useful, not Sufficient

Despite the ease with which technology controls can be circumvented, that is no reason to scrap the regime altogether. The MTCR can reduce the rate of technology transfer, and it- together with aggressive activity by the members of the Nuclear Suppliers Group--will slow the rate at which open and legal infrastructures for missile and nuclear weapon production can be built. But supply-side measures fail completely to address the central problem of nuclear and missile proliferation: *Nations attempt to procure advanced weapons because they perceive security problems which--they believe--can be ameliorated by possession of advanced weapons, because they harbor regional hegemonic ambitions, or because they believe that possession of high-technology weapons confers prestige. So long as those perceptions exist, and so long as the perception that advanced weapons improve security continues, supply-side controls will be insufficient to contain proliferation.*

In addition, supply-side controls nurture an underground economy which rewards precisely those nations, firms, and individuals who assist proliferators and who act in opposition to the general good of the world. The management of missile and nuclear proliferation must contain at least one other element: a carrot to encourage good behavior to supplement the bludgeon of controls on technology transfers. One example of a carrot would be cheap or free access to space launch services or space launch technology[10]. In the remainder of this paper I will address the question of whether it is technically possible to differentiate between missile and space-launch activities and technology with some confidence, leaving the issue of the political and economic feasibility of such a regime to other analysts. I will concentrate, instead, on whether compliance with a proliferation-management regime which permitted the development of space launch vehicles could be monitored.

Distinguishing Between Missile and Space Activities

On the surface there appears to be little distinction between the facilities needed to build missiles and space boosters. An almost equivalent statement is that there appear to be few differences between the products which are generally described as either ballistic missiles or space boosters. In fact, although the technologies for missiles and space launchers do have much in common, there are, in SALT II terms, "functionally-related observable differences" between a space-launch program and a program to build long-range ballistic missiles. These differences are readily noted upon close observation of the specific devices involved and in the way in which they are developed, tested, and deployed operationally.

Production Rates and Plant Capacity

Space programs, particularly in their infancy, are characterized by a small number of launches a year, and by a production rate which is roughly in balance with the launch rate. That is, a nation beginning a space program is unlikely to construct large numbers of boosters and to store them, because that is not economical, and also because one would expect space boosters to be fairly versatile rockets, adaptable to a wide range of payloads. Ballistic missiles, in contrast, need to be built in larger quantities, stored, and deployed with military units. Furthermore, ballistic missiles are apt to be much more standardized from unit to unit than are SLVs.

Thus, one may examine the production capacity a new space-faring or missile-proliferating nation constructs and match that capacity to the observed rate of flight tests (or space launches). Without attempting to propound specific rates at which a developing nation might launch payloads into space, it is, nonetheless, reasonable to suggest that even nations with the industrial base of the US launch only a handful of scientific satellites a year, and that a nation such as India constructs at most one or two scientific (including remote sensing) satellites per year. A launch rate of even half a dozen "scientific" payloads from a nation such as India or Argentina would be cause for concern that missile *development* and *flight testing* were occurring under the guise of space development. That, alone, would almost be sufficient to characterize the operation as military in character.

The first observable difference between a military production capability and a space launch capability is not the absolute rate but the balance between construction of airframes and propulsion systems on the one hand, and actual launches on the other hand. Although it is reasonable to suppose that there might be an accumulation of one or two unexpended boosters in a space program, an accumulation of ten would be cause for concern, and one of 25 would be grounds for active suspicion.[11] *Production rates or production capacities which significantly exceed observed launch rates of space payloads are strong indicators that the program is intended to produce ballistic missiles as well as space boosters.*

It is possible to estimate production capacity using well-established techniques. The most important of these is to measure directly the floor space available for constructing large rockets (or the first stage of a multi-stage rocket). Engineers with experience in industrial processes can readily estimate the maximum through-put of a rocket manufacturing facility if they know the size of the airframe and the area and layout of the factory. The launch rate is directly observable from the ground or from space.

Propellants

The designer of a booster intended to carry payloads into space will almost certainly turn to cryogenic propellants, either liquid oxygen (LOX) and hydrogen or LOX and a petroleum product such as kerosene. For space applications these propellants have the distinct advantage of providing significantly higher exhaust velocities, c^*, (sometimes measured in terms of the specific impulse, I_{sp}, defined as the number of pounds of thrust produced per pound of fuel burned per second. I_{sp}, measured in seconds, multiplied by the acceleration of gravity, 32.2 feet per second, gives exhaust velocity in feet per second).[12] Cryogenic liquids deliver higher specific impulses than do any storable liquid propellants, and vastly higher performance than any solid propellants. Their very nature, however, makes them much less suitable for military applications. The only cryogenic-fuelled missiles which were deployed by the United States (the Atlas ICBM, the Thor and Jupiter IRBMs, and the Redstone medium range missile) were all designed in the 1950s and retired by the 1960s. Descendants of the Thor and Atlas survive today, but only as the Delta and Atlas space boosters. From the operational point of view it is difficult to transport cryogenic liquids, particularly since their inevitable evaporation additionally stresses the logistical system.[13]

Therefore, it is reasonable to conclude that the development of cryogenic boosters poses less of a threat as a weapons program than would the development of boosters using either solid propellant or storable liquids such as nitrogen tetroxide and unsymmetrical dimethyl hydrazine (UDMH), a relatively high-performance combination.

High altitude sounding rockets pose a somewhat different challenge to the arms control regime. Because sounding rockets are generally quite transportable--indeed, one goal of the sounding rocket designer is to provide an efficient way to reach high altitudes above remote sites at very low cost--and do not need quite as high a specific impulse as rockets designed to place payloads in orbit, either storable liquids such as red fuming nitric acid (RFNA) and UDMH or solid propellants are suitable. However, the size, performance, and payload range of most sounding rockets marks them as potential testbeds for artillery rockets with ranges on the order of 100-150km rather than as prototypes of tactical ballistic missiles.

The propellants used in a large rocket can be determined unambiguously using remote sensing techniques.

Test Procedures

An end-to-end test of a ballistic missile differs greatly from a flight test of a space launch vehicle (SLV). Simplistically, the missile is expected to go up and come down at a distant point with an intact payload surviving the stresses of reentry (which need not be great if the range of the missile is less than around 1,000km); the SLV, in contrast, is expected to achieve orbital velocity with a velocity vector roughly tangent to the surface of the earth beneath. Although a space launch can be an acceptable surrogate flight path for the propulsion and guidance sections of an ICBM, it provides no information on reentry performance which includes RV (or airframe in the case of integral missiles such as the Scud B and *Al Husayn*) stability, heat shield survival, and the ability of electronics, etc., to function during and after the reentry process. In order to avoid unpleasant surprises on reentry such as those which plagued the *Al Husayn* in the Gulf War, it will be essential for any missile builder to conduct end-to-end flight tests which may even culminate with the detonation of an explosive at the target in order to confirm the functioning of the warhead.

Sophisticated conventional payloads, such as those which provide early dispensing of cluster munitions for chemicals or high explosives, clearly must be tested.

A reasonable clause in an accord which regulated missile proliferation might prohibit the testing of any rocket launched to greater than some specified range from point-to-point on the earth where the payload returned to earth intact other than by the use of a parachute. This approach would at least prevent or slow the development of heat shield technology. Another useful clause would forbid the flight of sounding rockets which were boosted *downward*, regardless of their point-to-point range. Such methods were frequently used in the early development of heatshield technology in the United States.

Each type of testing has a distinct set of signatures which is readily observable remotely, although not necessarily from convenient platforms in space. The specifics of such monitoring as carried out by, for example, the United States, are naturally classified, but the physics involved is sufficiently fundamental that the process can be described in general terms. *In summary, space launch vehicles can be distinguished from ballistic missiles by examining closely the operational procedures employed for their production and launch. This monitoring is more easily accomplished within the framework of a coopera- tive agreement than otherwise, but such an accord is not an absolute requirement.* Relevant verification and monitoring techniques will be discussed in the next section.

Verification and Monitoring of Missile Proliferation

Since controls on the export of missile technology from the existing missile powers are unlikely to prevent a proliferant nation from achieving its goals, and since monitoring of the flow of technology may not be possible, verification of the size, goals, and achievements of a missile program must rest on two fundamentally different sources of information: conventional espionage and technical collection.[14] These two data streams are complementary, and rely heavily on informed and technically expert interpreters working together. It is all too easy to forget that the British did not correctly interpret the information they had about the A4 until a test missile wandered off course and landed in Sweden, providing access to a nearly complete airframe and propulsion unit.[15] Until that moment the British estimates of the size of the missile varied by more than a factor of two, and supposed technical experts asserted that it must be powered by a propellant based on smokeless powder (cordite).

That said, human intelligence provided important clues to the nature of the A4 program, and a guess by Dr. R.V. Jones[16] led to the first identification of a missile on its *Meillerwagen* (a simple transporter-erector much like the improvised vehicles used by the Iraqis).

Verification and Monitoring Contrasted

It is appropriate to draw a sharp distinction between "*verification,*" which is the process by which nations assure themselves that their treaty partners have complied with their obligations, and "*monitoring*" which is the process of collecting information on the activities of another nation, whether or not such activities are covered by a treaty. *Verification,* by its very nature, can only take place within the context of a formal treaty or other form of agreed arms control accord. One *cannot* speak of verification when the process is purely one of monitoring or intelligence collection. Of course, monitoring is a crucial ingredient in amassing the data needed for verification. A discussion of the verification of missile capabilities ought to cover the more general question of monitoring techniques while still including those options available for providing adequate verification in a cooperative environment. In a cooperative environment, on-site inspections, for example, could be negotiated; in the absence of an accord they would probably not be. However, virtually all of the remote monitoring schemes would function as well without an accord as they would with one.

Monitoring Techniques in the Absence of a Treaty

Human Intelligence. HUMINT is, of course, a polite word for intelligence derived from spies and defectors; it can also include information from open source publications and from the reports of military attaches and other official visitors. Information from the "black" side of the intelligence community is generally considered to be reliable only when the source has proven credibility over a relatively long period.

Any HUMINT relies on the source having good access to the target information, and this access may vary over time. During the Second World War the Allies had relatively good information on activities at the Peenemünde development station, but these reports were pieced together from prisoner of war stories, tips from German scientists visiting in Norway, and from interrogation of captured Germans.[17] A source in place is clearly better, but more difficult to arrange.

A particular advantage of an agent in place is that reports may include microfilm of critical documents and intelligent analyses of what she sees or hears. In addition, a spy may have the opportunity to provide information about plans for the future and about intentions; neither plans nor intentions show up on satellite imagery. A disadvantage of intelligence derived from espionage is that the service employing the agent can never be certain whether or not the agent has been "doubled" or identified. In either of those situations, the information delivered by the agent will be precisely what the nation under study wants to have reported back and may contain various admixtures of truth, falsehood, distortions, and exaggerations.

Attache reports, on the other hand, can be believed since they come from a (presumably) trustworthy officer of the verifying intelligence service. However, it is unlikely that a military attache or science counselor will ever be permitted access to any information which the monitored state desires to conceal. The possibility of an attache being fed disinformation also exists, and "Potemkin Missiles" should not be excluded.[18] *Although spies may not be essential for verification and monitoring, they make the job much easier and can provide information on plans and intentions in addition to technical capabilities, which can often be determined using technical verification and monitoring methods.*

Optical tracking. When the launching facility is located close to a coast line or to an international border with a cooperative nation, ships or aircraft can be positioned to track and photograph missiles in flight. If a test vehicle--SLV or missile--is expected to leave the boundaries of the launching nation, it is customary to announce the fact in advance through a "Notice to Mariners" and a "Notice to Airmen;" such notifications are one way in which monitoring assets can receive the necessary cues to move into position.

Optical tracking of a missile from a satellite in low earth orbit (LEO) is impractical unless the launching state is particularly cooperative. This is because the orbits of LEO monitoring satellites seem to be easy to recognize, and the launching state can easily select launch times when no LEO assets are in position. The coverage of a conventional reconnaissance satellite is, in any case, limited to the ability to snap a "photo" of the test site, perhaps with the rocket in flight. In general one would not expect such a satellite to have the ability to maneuver its camera so as to follow a test vehicle in flight.

Not all optical tracking information of relevance comes from images of the tested vehicle. A great deal of useful information can be extracted from studies of the rocket exhaust. In general one may suggest that the spectrum of the exhaust should serve to identify the propellants and, if the spectrum is detailed enough to determine the reaction products and their proportions, something about the efficiency with which the motor performs. The absolute brightness of the exhaust (most of which is in the thermal infrared) can provide an indication of the rate at which fuel is burned, and hence the thrust of the rocket. In turn, this can be used to infer throwweight over a given range.

If a treaty permitted on-site tracking or inspection, cooperative photography might be extremely useful; payload inspection and careful inspection of the putative SLV's nose shroud would be more definitive.

Imaging Data Obtained from LEO

Conventional Imagery. Ballistic missiles with ranges great enough to be of concern to the non-proliferation community tend to be large enough to be detected on imagery with a resolution[19] as poor as 5 meters and identified on imagery with resolution of 2 meters or better. The V-2 was approximately 13 meters long; while it could have been detected at the 5 meter level (not fewer than 3 pixels would have been affected, and as many as 7 could have been if the object's placement were precisely correct), this would *not* have been sufficient evidence to assert the identification of a rocket. At a resolution of 2 meters a V-2 would have filled 15 or 16 pixels and would have had a distinctive appearance; if one meter resolution were available, the fins of the rocket would have been clearly identifiable, and significant information about the missile could have been gleaned by an analyst familiar with rocket technology.[20]

Missiles of this size are, however, transportable and concealable. With any understanding of the operation of the observation satellite system, the missile builders could take perfectly adequate steps to conceal missiles from view with only a minimal impact upon their programs.

From the point of view of verification and monitoring, the bigger and the more single-purpose an installation is, the easier it is to detect and identify with confidence. It also becomes far harder to conceal. For example, the author has positively identified the Groom Lake Air Force installation in Nevada, where the "stealth" (F-117) fighter aircraft was tested, using only Landsat imagery. It was possible to measure the runway lengths and widths and to make a reasonable inventory of the size distribution of hangar- and factory-like buildings at the site. No aircraft could be detected unambiguously.

Remotely sensed data useful for verification and monitoring comes, fundamentally, from three sources: the reflected visible and infrared spectra (called "optical" data), emitted long wavelength infrared radiation (called "thermal" data because the amount of energy radiated by an object is a function of its temperature), and reflected microwave energy from a radar satellite. Normally the radar data must originate from a synthetic aperture system because the great wavelength of microwaves (centimeters, not nanometers) would degrade the resolution of the system unless a large receiving antenna could be synthesized from one of practical size by the motion of the satellite.

Our eyes have evolved to display the world in color because the additional information is useful for our survival. Similarly, imagery is evolving from the black and white film and electronic imagery used in earlier days of photoreconnaissance towards multispectral pictures which break the spectrum of light reflected from a target into many wavebands ranging from visible blue through the near- and mid-infrared. The information provided by such "color" data permits greater discrimination between real targets and decoys and provides an opportunity to detect--and sometimes penetrate--camouflage. The Landsat 4 and 5 satellites are examples of civilian applications of the techniques of multispectral imagery which can be computer processed to classify different structures and terrain so that tiny differences are immediately visible to the eye.

Each distinct area in a multispectral image is likely to display a typical combination of reflectances in each of the wave bands--a *relative* distribution of spatial energy so that shadowed areas will have similar signatures to areas in full sun--and the analyst can seek clusters of reflectance values in a multidimensional space. Pixels belonging to individual clusters can be assigned color values which are then displayed. Classes can be assigned automatically by the computer (unsupervised classification) or by the analyst based on information known about parts of the scene (supervised classification).[21] Although classification must be used with care since the vector space spanned by the finite number of wave bands with finite spatial and spectral resolution is not complete (meaning that similar signatures do not *necessarily* represent identical terrain or features on the ground), both supervised and unsupervised classification are among the most powerful instruments in the verification toolkit.

Thermal IR from LEO. In the author's experience thermal imagery is the hardest type of conventional picture to deal with. Thermal IR data typically

Table 11.1 IFOVs Required for Recognizing Various Targets Associated With Ballistic Missile Verification

Targets	Detection	Recognition	Technical description
short-range	2.5 m	.5 m	.3 m
medium-range	3-5 m	1 m	.5 m
IRBMs (e.g. PII)	5 m	2-2.5 m	.5 m
ICBMs	5-10 m	5 m	1-2 m
Factories	30 m	20 m	5-10 m
Launch sites (IRBM)	30 m	10 m	5 m
Test ranges	>80 m[b]	80 m	10-30 m
Static test facilities	30 m	10 m	1-2 m
Associated military facilities	>30 m	30 m	10 m

[a] This table is based on the author's personal experience examining satellite imagery from US, French, and Russian **unclassified** sources as well as his experience with World War II aerial photographs of the facilities at Peenemünde and elsewhere in Germany and his inspection of the few publicly-released U-2 photographs of former Soviet facilities taken in the late 1950s. As used here "detection" means the unambiguous detection that something of the *general* character of the target is present on the ground; "recognition" means that the character can be specified (e.g. ICBM); and "technical description" means that the analyst can name the missilie (e.g. SS-19 ICBM) or describe a facility in detail. Note that identifying a specific missile type is only possible if the same missile type has been seen before and given a name. For these IFOVs to be relevant, a missile must be seen lying horizontally. [b] Several Soviet test ranges were identified using the Landsat 1,2, and 3 "MSS" sensors with resolutions of 80 meters. These pictures were published in *Aviation Week* in the 1970s

range in wavelength between 8.5m and 12.5m, and so are difficult to focus resulting in poor spatial resolution. In addition, the power received at the satellite is low, and the background of thermal noise from the electronics requires the sensors to be cooled at least to liquid nitrogen temperatures, if not below. Finally, the amount of energy radiated by an object depends not only on its absolute temperature, but also on its emissivity, a quantity which is a function of the composition and surface texture of the object under study.[22] Hence, not fewer than two colors of thermal IR must be detected and the absolute power levels at satellite measured before it is possible to estimate the temperature of an unknown surface, even in principle. This simple description omits difficulties which arise from absorption in airborne moisture and the near-impossible task of modelling the atmospheric absorptivity in the presence of humidity, various aerosols, and many possible pollutants.

Thermal IR is, therefore, most useful for detecting modest temperature differences between locations where the surface properties are well known (for example, in a lake or reactor cooling pond), but where the locations are not so distant from one another that the moisture content of the atmosphere is likely to differ greatly. With the single "color" thermal IR sensor ("band 6") aboard the Landsat 4 and 5 satellites it is possible to distinguish hot and cold areas of lakes and rivers, for example adjacent to the water outlet of a nuclear power plant.

With a two-color signal it would not be impossible to distinguish between areas of high and low activity within a factory so long as the spatial resolution of the system were significantly superior to that of the 120 m IFOV of the Landsat system.[23] Cryogenic facilities are not likely to have a significant thermal IR signature because their designers must minimize heat transport from the environment to the liquified gases; in consequence, the exterior of a liquefaction or cryostorage facility is apt to be no colder than the ambient temperature of ordinary buildings.

Radar Data from LEO. Synthetic aperture data from LEO can provide significant advantages to the verifier because SAR operates as well at night or through cloud cover as it does on a clear day. It also can cover a large swath width without suffering undue loss of resolution at the far edge of the image.

SAR also offers the additional advantage, since it operates at night, of doubling the number of potential observing opportunities as compared to visible and near-IR sensors which require solar illumination.

On the other hand, SAR images are significantly more complicated to interpret than ordinary images, whether in the visible or infrared wavebands because the energy received back from an object illuminated by radar depends upon a number of factors such as:

- electrical conductivity of the surface
- angle of the surface with respect to the incident radiation
- texture of the surface,

while the brightness of a surface in an optical image depends almost entirely on its reflectivity.

As an example of the first three factors in interpreting SAR images, water usually appears black in a radar image since the incident radiation were reflected away from the receiving antenna; although in a storm where the water is rough, it might appear gray or even fairly bright. Metal roofing could appear either black if the radar energy is reflected specularly away from the satellite, white if the angle is such as to send the energy back to the receiving antenna, or depending upon the

• regular periodic arrangement of electrically conductive objects and surfaces which might act as a large antenna,

corrugated metal roofs could even appear as brilliant point sources if they acted as directional antennas. Since SAR data are routinely used for many purposes, it is, of course, possible to train photoanalysts to use radar data correctly--but the additional training is imperative. Although I have worked extensively with visible and infrared imagery and am a reasonably competent analyst, I would not attempt to do a serious interpretation of a radar image.

In addition, it is well known that large amounts of computing power are needed to convert the SAR data to useful images. A final disadvantage of active methods of probing, including LIDAR (Light [or Laser] Detection and Ranging, the laser version of radar) as well as SAR, is that the act of probing is detectable by the observed state. A probed state can observe the SAR signals and deduce from them the satellite which carries the sensors and, with some difficulty, the exact coverage pattern and resolution of the instrument. This information detected on the ground can allow the development of concealment techniques and jamming. Detection is made simple because the power reflected to a radar antenna varies as the inverse *fourth* power of the distance between antenna and target; the detectable power on the ground varies only as the inverse *square* of the distance. Just this property permits cheap radar detectors in cars to defeat expensive radar guns used by the police.

Signals and Electronic Intelligence

During the SALT II negotiations one sticking point between the US and Soviet sides was the Soviet practice of encrypting telemetry from ICBMs to the ground. One might infer from the US position that it is possible for orbiting satellites or ground stations to receive telemetry meant for other destinations. One might also infer, from the discussions pertaining to encryption, that it might be possible for a receiving nation to unscramble ordinary telemetry to determine something about the performance of a rocket in flight. Any such capabilities must be among the most sensitive issues of verification and monitoring and cannot be discussed in an unclassified environment.

The laws of physics, however, offer some insight into the kinds of information which can be gleaned *without* recourse to the details of the telemetry, given only that the SLV or missile carries any sort of radio transmitter and that a suitably located receiver is available to track the signal. A single receiver, which can be located on the ground, at the geostationary orbit, or at a convenient location in between is capable of recording the "beacon" signal from the rocket. If the beacon is detected before launch, its fundamental frequency can be unambiguously determined.

Given knowledge of the fundamental frequency, the doppler effect permits the receiving station to determine the instantaneous component of velocity along the line between the missile's velocity and the receiver. By differentiating the measured velocity, instantaneous acceleration can also be determined. Three or more doppler receivers in known locations can be used to determine the total (scalar) velocity and (scalar) acceleration of the rocket. The actual flight trajectory can be reasonably well determined if the launching and impact points can be located.[24]

More information can be extracted from the doppler data. If one knows the entire acceleration vs time curve as well as the final velocity of a rocket, it is possible to infer something about the specific impulse of the propellant used, which can be seen by differentiating the "rocket equation" with respect to time:

$$v(t) = c^* \times \ln(m_o/m(t))$$

where $v(t)$ is the instantaneous velocity, c^* is the exhaust velocity, $m(t)$ the instantaneous mass, and m_o the take-off mass of the rocket.[25] With this information it is possible to infer something about the launch weight and throw weight of the rocket if it is used as a missile.

Characterization of the I_{sp} of the propellants being used is extremely useful in deciding whether they are of a sort which might be used in a military rocket or only in a research vehicle.

An alternative use of signals intelligence is as a cuing device. That is, radio messages sent up and down a test range can provide significant information about what kind of rocket is to be launched, when, and to what distance and at what azimuth. Even if the traffic is encrypted, the mere presence of additional activity plus the turning-on or testing of range instrumentation provides useful clues which can assist in positioning or tuning of specific monitoring and verification assets. Receiving stations for SIGINT and its companion ELINT (Electronics Intelligence) can be land, sea, air, or space-based. When a treaty requires advance notification of testing, SIGINT provides a means to verify the likely accuracy of a nation's notifications and also provides the opportunity to position verification assets in advance of a test.

Finally, SIGINT provides another check on whether or not a nation is developing the capability to use long-range ballistic missiles. Military organizations cannot function without doctrine, training, and deployed equipment. Although a nation might succeed in concealing the size of its arsenal, it is doubtful if it could be confident in its ability to employ ballistic missiles unless its troops trained with them and unless it exercised its C^3 (Command, Control, and Communications) system for missile use on a regular basis. Most of this required training and C^3 development must take place by radio, and hence the messages can be intercepted. It may not be necessary to decipher encrypted traffic if a particular pattern of traffic can be associated with a specific set of activities related to missile launches.

Non-imaging Optical Data

Because much of the powered flight path of a missile takes place far above the sensible atmosphere, it is possible to use a satellite to detect wavelengths in the missile's exhaust plume which would be absorbed by the atmosphere. These include mid-IR and ultraviolet. Plume tracking can provide detailed information on the chemical composition of the propellants in use through their emission and fluorescence properties. Broadly speaking, the emission spectrum in the infrared provides a fingerprint of the chemical species in the reaction products of the propulsion system as described earlier in relation to optical tracking. The additional information which may be gleaned from detecting the radiation emitted by compounds which fluoresce under the sun's short wavelength UV is also useful, as is the actual UV emission spectrum of the plume. In 1986 the defense press reported that the "Delta 180" experiment performed by the Strategic Defense Initiative Organization (SDIO) carried UV detectors. *Technical means which are deducible from the laws of physics can be readily discussed in the open literature, and provide insights into useful techniques for monitoring missile proliferation. The details of such systems must remain classified. All technical collection benefits from supplementary HUMINT.*

Verification in a Cooperative Environment

Most of the techniques described above are as applicable to a cooperative environment as they are to the kinds of verification written into the arms control treaties of the SALT and START eras when the possibility of obtaining the cooperation of the monitored party approached zero. In a treaty-regulated setting, where all parties have agreed to limits on missile development and on isolating space launcher technology from missile technology some additional steps may be taken. These include the obvious ones of stationing inspectors at rocket test ranges to verify that the payloads are of a scientific, commercial, intelligence gathering, or other permitted type and that reentry vehicles have not been fitted to the missile. In addition, plume analysis can be simplified and raised to a higher degree of accuracy if sensing instruments are located near the launch point and beneath the early trajectory of the rocket. These can be enclosed in "black boxes" or operated by inspectors.

In any event, quantitative data streams from cooperative monitoring equipment must be secure and authenticated and available to both parties. Even in the most cooperative environment imaginable, one must anticipate that the monitored party will still insist upon knowing what information is being transmitted from the sensing equipment.

A further possible advance for verification in a cooperative environment would be the installation of monitoring equipment (operated by inspectors on-site or remotely) at critical production facilities. It is conceivable that treaty-limited items could be tagged, inventoried, and followed from manufacture through launch. One might even envision obtaining access to the software and firmware used in the guidance systems of space launch vehicles to ensure that the same guidance systems could not be quickly converted to the control of a long-range ballistic missile. *A cooperative monitoring regime is superior to a non-cooperative one, but on-site inspection is no panacea. One must always keep in mind Fred Eimer's First Law: "A violator will never let you get to the scene of a violation before the site has been cleaned up so that no violation is apparent."*[26]

Verification Without the United States

At present the American capacity for technical verification far outstrips that of any other nation; indeed, it may outstrip the capabilities of all other nations combined. Nonetheless, in a cooperative environment there is no reason to believe that the full power of the US intelligence community is needed. Certainly almost any nation of Western or Eastern Europe and many nations of the Far East can design and produce monitoring instruments designed for emplacement on-site. The *Helios* reconnaissance satellite being developed by a French-Spanish-Italian consortium for launch in 1994-95 reportedly will have a one meter resolution and is adequate for virtually all of the tasks in connection with medium range missiles as listed in Table x.1, which is empirically-based. Existing civil remote sensing satellites such as the French *Spot* can certainly be used to measure floor space in factories, observe major activity at launch sites, and provide information which can be used to designate sites for on-site inspection. *Spot* and Landsat were used for precisely this purpose by the United Nations Special Commission on Iraq.[27]

When it comes to so-called "national technical means of verification," fundamentally imaging and other intelligence-gathering satellites, there are some substitutes for resolution. These include frequency of coverage, spectral range and resolution, and a well-concealed orbital pattern. Unfortunately, none of these come cheaply, and none are as close to the grasp of a developing country's technology as are the weapons themselves. It is difficult to envision a "bronze medal" in verification being possible, or worth very much, since it is likely to be easier to cheat than to catch a cheat at the lower end of the military-technical scale.

Conclusions

Restrictions on the sales of high-tech products with both direct and indirect application to missile and nuclear weapon development will not obviously harm the fundamental effort to retard the spread of such weapons, but will unnecessarily handicap the industries of the United States which must compete with other nations for sales in a bizarre bazaar. For that reason, most export restrictions ought to be eased. This assertion is particularly true of computers, including supercomputers.[28] Even so, by slowing down the rate at which developing nations can acquire new technologies, the MTCR and similar efforts do raise the political and financial costs of proliferation. At the same time, the apparent discrimination against emerging nations embodied in non-proliferation regimes makes acquisition of high-status weapons--those the great powers have and wish to deny to the rest of the globe--more attractive.

Supply-side controls are also likely to fail because the bronze medal technologies needed for an entrance ticket into the military-technical Olympics are widespread. They have become integral parts of modern industrial and silicon society, and cannot be limited severely without condemning the non-recipient nations to third-class industrialization and a perpetually lower standard of living. Indeed, the developing world already includes mid-level nations which can supply most of the technologies in question to the nations just beneath them on the ladder. A broad commerce in missile and nuclear technologies already exists outside the bounds of the MTCR participants and the Nuclear Suppliers Group.

By themselves supply-side controls will not accomplish the task of non-proliferation, for nations do not invest in expensive weaponry for its own sake. Ambition, status, perceived threats to security, and a desire to dominate a region are greater spurs to the proliferation of advanced weapons than is the ready availability of useful technology. In turn, the difficulty of achieving success with unilateral controls leads to the recognition of the desirability of reaching agreement on limiting or managing the proliferation of ballistic (and cruise) missiles with ranges in excess of about 300km. If such a treaty could be achieved, it is technically possible to distinguish between peaceful activity in space exploration and exploitation based on certain characteristics of the rocket-building and rocket-testing complex as well as by observing military training and exercises. For the most part, verification can be achieved using various remote sensing tools ranging from imagery to signals intelligence. The ability to avoid intrusive verification methods may make such an accord more attractive to nations which could easily have dual capabilities-- SLVs and long-range ballistic missiles--but are willing to surrender their missiles.

Notes

1. The opinions expressed in this article are solely those of the author and not necessarily those of the Center for Strategic & International Studies or of any organization which has funded the author's research. The author thanks the Carnegie Corporation of New York for a discretionary grant received while he was at the George Washington University which permitted the analysis on which the formulation of the bronze medal idea is based.

2. See P.D. Zimmerman, "Bronze Medal Technology: A Successful Route to Nuclear and Missile Proliferation," *Orbis* (Winter 1994), for a more detailed analysis of the processes by which a nation can acquire the minimalist technology needed for a satisfactory "bronze medal" level missile or nuclear program.

3. See, for example, P.D. Zimmerman, *Iraq's Nuclear Program: Sources, Achievements and Stature*, Congressional Research Service, February 1993, for a full account of a "bronze medal" program in weapons of mass destruction which was nearly successful and which relied to an astounding extent on indigenous technology, in particular for the entire explosive assembly including switching devices, capacitors and detonators.

4. According to Dr. Waldo Stumpf, CEO of the South African Atomic Energy corporation in a talk delivered at the South African Embassy, Washington, DC, July 23, 1993.

5. The mean accuracy of the Fi-103 is given as 6km in range and 5.8km in azimuth by captured German documents. These values resulted from a series of 189 flights of missiles drawn from the series production over ranges from 175 to 225km. The data showed a range bias of +0.6km and a lateral bias of 2.6km from the aim point. See *Bericht über Leistungen, Treffgenauigkeit, Versagerursachen beim Schleudereinsatz FZG 76 auf Grund der Erprobung mit Großseriengeräten bis zum 15.8.1944*, p. 7 (hereinafter cited as *Leistungen, Treffgenauigkeit*). This document, in the archives of the Smithsonian Air and Space Museum, originated from *Erprobungsgruppe Temme*, Karlshagen, Germany and carries the German classification *Geheime Kommandosache*. In

some German references, including this one, the Fi-103 is referred to by the code name FZG 76 (*Flakzielgerät 76* or anti-aircraft target apparatus 76).

6. From a test series of 265 missiles, only 64 Fi-103s failed to reach their target area. *Leistungen, Treffgenauigkeit*, p. 9.

7. David Kay, *Eye on Supply*, Monterey Institute, Spring 1993.

8. The longer range is of greater military usefulness as a glance at a globe will show. A range of 10,000km spans precisely one fourth of the earth's circumference and permits virtually any country with long-range rockets to reach any targets within its geopolitical adversaries.

9. Overpressures from nuclear explosions are usually given in pounds per square inch, if only because the only openly available data come from American publications describing engineering experiments conducted with atmospheric nuclear explosions in the mid-1950s. An overpressure of 5 psi is sufficient to destroy virtually all residential construction and most commercial or industrial structures. Thus, for the purpose of estimating the requirements to impose on the planners of a third world nuclear ballistic missile strike force, target overpressures of 5 psi are generally adequate if the doctrine for employing the weapons is fundamentally countervalue (populations) or countermilitary (deployed troops).

10. From the point of view of slowing or managing proliferation, access to launch services is preferable to access to launch technology, but this restriction may not be acceptable to all potential proliferants.

11. The numbers "10" and "25" are drawn from the author's experience with what is reasonable to expect and from a comparison of the production rates of US space boosters compared with the production rates of US ballistic missiles. The specific figures of 10 and 25 should be taken as *illustrative* and not as definitive trigger quantities.

12. A high exhaust velocity is critically important to space missions, because the final velocity which the rocket can achieve varies linearly with c^* but increases only logarithmically with the ratio of the take-off mass of the vehicle to its payload mass. The designer must work very hard to increase the mass ratio, which brings only small rewards; a 10 percent increase in exhaust velocity brings a direct 10 percent increase in final velocity and a 20 percent increase in range.

13. The V-2 is, of course, the notable operational exception of a mobile missile which used cryogenic propellants. Since no other propellants were available to its designers, Von Braun and his team, LOX was used as the oxidizer and grain alcohol was the fuel. Liquid oxygen and its transport formed one of the major bottlenecks in the missile's development and operation.

14. For another view on verification of missile proliferation, see Ruth H. Howes, "Monitoring the Capabilities of Third World Ballistic Missiles" in *Missile Proliferation, Missile Defense, and Arms Control*, Götz Neuneck and Otfried Ischebeck, editors, Nomos Verlagsgesellschaft (Baden-Baden, Germany: 1993), pp 101-113.

15. R.V. Jones, *Most Secret War: British Scientific Intelligence 1939-1945*, Cornet Edition, Hoder and Stoughton Paperbacks (Dunton Green, Sevenoaks, England: 1979), p. 544. Note: this book has appeared with slightly varying titles and under a number of publishers' imprints.

16. Ibid., p. 434.

17. Arnold Kramish, *The Griffin*, Houghton Mifflin Company (Boston: 1986), Chapter 13, p. 63.

18. The most famous such disinformation was probably provided to the Eisenhower administration by the Soviet government. It is widely known that at one May Day parade the Soviets took their one small group of jet-powered heavy bombers and flew them repeatedly over the parade route, thus causing the Western attaches in attendance to report that the Soviet Union possessed a very capable fleet of nuclear bombers. Thus began the famous "bomber gap."

19. "Resolution" as used here means the instantaneous field of view (IFOV) or pixel size of an electro-optical camera. In order to convert IFOV to photographic resolution given in terms of meters per line pair, one must multiply the IFOV by a number generally accepted as being between 2.25 and 2.75. As a rule of thumb, 2.5 is satisfactory.

20. See P.D. Zimmerman, *Using Synthesized Images to Establish Monitoring Capabilities*, Volume 68 in the series *Hamburger Beiträge zur Friedensforschung und Sicherheitspolitik* (Hamburg Contributions to Peace Research and Security Policy), Institut für Friedensforschung und Sicherheitspolitik, Hamburg University, October 1992 for synthetic digital images of various kinds of military hardware in order to obtain an idea of the resolution requirements for performing various tasks.

21. Many books cover classification techniques in detail. A good introduction is to be found in Floyd F. Sabins, *Remote Sensing Principles and Interpretation*, 2nd ed., W.H. Freeman and Company (New York: 1987). See the index for the appropriate sections.

22. Counterintuitively, the more reflective an object is, the *lower* its emissivity, and the darker it is, the *higher* the emissivity. Black bodies have an emissivity of 1.0; shiny aluminum has an emissivity as low as 0.1 depending upon the alloy and the surface treatment.

23. Landsat thermal IR data were used to study the cooling pond of the Chernobyl reactor complex after Unit 4 burned in April 1986. See Frank G. Sadowski and Steven J. Covington, *Processing and Analysis of Commercial Satellite Image Data of the Nuclear Accident Near Chernobyl, USSR*, US Geological Survey Bulletin 1785, 1987.

24. Note that doppler information is *only* available if the rocket being tested transmits a radio signal of some sort. A nation which decided that it would cheat on an obligation not to conduct certain kinds of flight tests might elect not to transmit any radio signal from its test vehicle and simply record all information in a recovery package. There is precedent for conducting flight tests in this manner. Note, as well, that a surface-to-surface test over an extended range could itself be a banned activity under a missile non-proliferation accord.

25. Interested readers will carry out the differentiation and remember that c^* is I_{sp} times the acceleration of gravity. Thrust is, of course, $c^* \times dm(t)/dt$ for a stationary rocket, but for an accelerating vehicle the thrust is equal to the time derivative of momentum.

26. Dr. Manfred Eimer was Assistant Director of the United States Arms Control and Disarmament Agency for nearly eight years under Presidents Reagan and Bush.

27. At the request of the IAEA I used Landsat imagery to examine a possible site of an underground nuclear reactor in Iraq. The fact that no reactor was located there does not change the fact that Landsat was useful in assessing the major changes in the area and in finding suspicious sites within the general region.

28. An 80486-based machine is more powerful than any computer that existed when the first compact hydrogen bombs and the first ballistic missiles were designed and built. During the Manhattan Project "computer" was a job title, not a machine. It is not wholly unreasonable to suggest that the progress of some entry-level proliferants might be retarded by a supercomputer. That is, it might take so long to master the sophisticated programming languages needed to exploit supercomputer power that the task might be completed more rapidly on a desktop. It is also possible that programmers assigned to a new supercomputer might spend so much time hacking for pleasure that they would be less efficient at programming for their jobs.

About the Contributors

Yanping Chen is a specialist in China's space program and policy at the Space Policy Institute and a specialist in China's science and technology programs and policy at the Center for International Science and Technology Policy, George Washington University. Before her career in the United States, she worked in China as a senior scientist in space medical research and a senior executive for China's man-flight program. Contact information: Space Policy Institute, Washington DC. Fax: 1 202 9941639.

Peter Hayes directs the Security Program at Nautilus Institute for Security and Sustainable Development in Berkeley, California. He is the author of *Pacific Powderkeg: American Nuclear Dilemmas in Korea* (1991) and co-author of *The Global Greenhouse Regime: Who Pays?* (1994) and *American Lake: Nuclear Power in the Pacific* (1987). Fax: 1 510 5269297; email: npr@igc.apc.org

Joan Johnson-Freese is on the faculty at the Air War College at Maxwell Air Force Base in Alabama. This paper was written while she was Director of the Center for Space Policy & Law, and an Associate Professor of Political Science, at the University of Central Florida. Her primary area of research is international cooperation and competition in space, and she has published on that topic in such journals as Space Policy, Spaceflight, Nature, Technology in Society and Space Commerce. She is also the author of *Changing Patterns of International Cooperation in Space* (1990) and *Over the Pacific: Japanese Space Policy into the 21st Century* (1993). Fax 1 334 9534208.

Lora Lumpe is a Senior Analyst with the Federation of American Scientists in Washington, DC. She directs the Federation's Arms Sales Monitoring Project, which she founded in 1990. The project promotes restraint in U.S. and global conventional arms production and trade. She edits a bi-monthly newsletter, the Arms Sales Monitor, which reports on and analyzes U.S. policy in this area. Ms. Lumpe has published in *Scientific American, the Bulletin of the Atomic Scientists,*

233

Arms Control Today, Arms Control, newspaper editorials and book chapters. Fax: 1 202 6751010; email: http://www.fas.org/pub/gen/fas/

Molly K. Macauley is a Senior Fellow at Resources for the Future (REF) in Washington, DC, and a professor in the Department of Economics at Johns Hopkins University. At REF she directs the research program on economics and policy issues of space. Dr. Macauley has published widely on the economic value of the geostationary orbit and electromagnetic spectrum, technical change in the communications satellite industry, and problems in space transportation, space-based remote sensing, and space debris. Contact information: Resources for the Future, Washington, DC. Fax: 1 202 9393460.

Thomas G. Mahnken is a Senior Analyst with SRS Technologies in Arlington, VA. He served as a member of the Secretary of the Air Force's Gulf War Air Power Survey examining the role of air power in the Gulf conflict and as an analyst in the Non-Proliferation Directorate, Department of Defense, where he monitored missile proliferation. Fax: 1 703 5222891.

Timothy V. McCarthy is Senior Analyst, Center for Nonproliferation Studies at Monterey Institute of International Studies, Monterey, California. He specializes in missile and nuclear proliferation. He is also on missile inspection teams for the United Nations Spcial Commission. This paper was written before McCarthy became an UNSCOM inspector. Fax: 1 408 6473519; e-mail: tmccarthy@miis.com

Janne E. Nolan is a Senior Fellow at the Brookings Institution. She is the author of *Trappings of Power: Ballistic Missiles in the Third World* (Washington: Brookings, 1992) and *Guardians of the Arsenal: The Politics of Nuclear Strategy* (New York: Basic Books, 1989). Fax: 1 202 797-6003.

John Pike is the Director of the Space Policy Project at the Federation of American Scientists. He coordinates the Federation's research and public education on space policy. A former political consultant and science writer, he is the author of over 170 studies and articles on space and national security, and is a co-author of the book *The Impact of US and Soviet Ballistic Missile Defense Programs on the ABM Treaty*. He advised the 1984 Mondale campaign, the 1988 Dukakis campaign, and the 1992 Clinton campaign on defense and space policy issues. Fax: 1 202 675-1010.

Jürgen Scheffran is a member of the Interdisciplinary Reserach Group in Science, Technology, and Security Policy at the Technical University of Darmstadt, Germany. He has written extensively on missile and space technologies and verification approaches. Dr. Scheffran is a member of the faculty in Physics. Fax: 49-6151-164321.

M. Lucy Stojak is on the faculty at the Centre for Research in Air and Space Law at McGill University in Montreal and at the International Space University. Fax: 1 514 9376559.

Eric Stambler is a Research Analyst with the Space Policy Project of the Federation of American Scientists. Fax: 1 202 6751010.

Maxim V. Tarasenko is a space programs analyst at the Center for Arms Control, Energy and Environmental Studies at Moscow Institute of Physics and Technology, Russia. This paper was written while he was a Visiting Fellow at *The Bulletin of the Atomic Scientists*. Fax: 7 095 4085336; Email: Tarasenko@glas.apc.org

Peter Zimmerman is a Visiting Senior Fellow at the Center for Strategic and International Studies in Washington, DC. Fax: 1 202 775-3199.

About Nautilus Institute

Nautilus Institute is a policy-oriented research and education organization that promotes international cooperation for security and ecologically sustainable development. Programs embrace both global and regional issues, with a focus on Asia-Pacific. Nautilus Institute produces reports, organizes seminars, and provides educational materials and training services for policymakers, media, researchers and community groups.

Core staff are based in Berkeley and Washington, with associates in Tokyo, Seoul, Shanghai and Moscow. Research draws from many disciplines, including environmental economics, natural sciences, energy and resource planning, and international relations.

Nautilus Institute is a non-profit organization funded primarily by grants and contributions. The Institute can be contacted at:

1831 2nd St,
Berkeley, CA 94710
USA
ph 1 510 204 9296
fax 1 510 204 9298
email: npr@igc.apc.org
WWW: http://www.nautilus.org/nautilus

About the Book

In this unique volume, an international cast of leading scholars from several disciplines offers a comprehensive assessment of the current status of space-based weaponry. Regional and technical experts offer their analysis of the major powers' special interests in space and also examine the broader issues of ICBM proliferation, testing, monitoring, and verification as well as possible opportunities for cooperation between states with a stake in space power.